Praise for Michael E. Gerber, Ken Goodrich, and *The E-Myth HVAC Contractor*

Michael E. Gerber will tell you what you need to do. **But it's Ken Goodrich that will tell you exactly how he applied those insights in the world of HVAC to create a series of legendary successes.** Has anyone ever made as much money as Ken Goodrich in HVAC without rolling up a series of companies built by other people? Ken isn't a consolidator of other people's companies. He's an operator of his own companies. And he's got a world of powerful things to teach you if you're willing to learn.

Roy H. Williams, *New York Times and Wall Street Journal*
author of *The Wizard of Ads* trilogy of business books

Ken is one of the greatest and most successful entrepreneurs in the history of contracting. It's been a real privilege getting to see his success over the years. He started from the bottom and build his extraordinary company from the group up. **And now, he's giving back to the contracting community in a big way by sharing his story, his playbook, and his secrets to help others create success for themselves.**

Ara Mahdessian, CEO, ServiceTitan

In the nineteenth century, Horatio Alger wrote fiction about humble boys rising from rags to riches, which constitutes the great American success story. Ken Goodrich's story could have been written by Horatio Alger, except that Ken's is not fiction. From his days of a mullet in a truck, Ken has risen to become America's most successful residential air conditioning contractor. In this book, he not only shares the lessons he's learned, but he teams up with the iconic author of *The E-Myth Revisited*™ and other great books for small business, Michael E. Gerber. **This is a must-read for any contractor, but especially for contractors living lives of sustenance in an industry that can deliver lives of prosperity.** Ken and Michael show you how.

Matt Michel, President, Service Nation Inc.

I have known Ken for fifteen years, as a father (our kids were similar ages), husband (our wives were friends), business partner, and friend. He is the real deal—hardworking, sincere, ethical, thoughtful, and action-oriented. Few leaders listen like Ken. He is constantly seeking to learn and to improve himself and his business. He believes that success is a team sport and ensures that others prosper when he does. Ken believes that work should be fun! **He consistently builds high-performance organizations where each member of the team can make a difference and enjoy the winning culture.**

Richard Haddrill, CEO, The Groop, LLC;
Vice Chairman, Scientific Games, Inc.

Ken is a results-oriented entrepreneur with a passion for doing things the right way. In his current business, Goettl Home Services, they live by a motto: "We do things the right way, not the easy way." In addition to doing things the right way, I am very impressed by Ken's boundless curiosity and not at all surprised that he finds inspiration in Michael E. Gerber and the Seven Centers of Management Attention. **If you want to build a successful business, Ken Goodrich and Michael E. Gerber are the winning combination.**

Jonathan E. Baum, Managing Partner, Baum Capital Partners

When I first read *The E-Myth Revisited*™ in 2000, my business and life changed forever. If only Ken's book had existed back then, I would've had the industry-specific insight to massively expedite the growth and profitability of my HVAC company. **Ken Goodrich is a true master of the HVAC business, and in this book he shares the powerful systems, strategies, and techniques that have helped him become one of the greatest operators the industry has every seen.** Do yourself, your team, your customers and your family a big, big favor; immerse yourself in this information, implement the strategies, and enjoy your next level of success!

Kenny Chapman, founder, The Blue Collar Success Group

Ken and I shared paths that are very close to the same. Both of us started companies and struggled until it was almost time to give up. Then our search for the answers brought us to Michael E. Gerber's E-Myth. We were half a country apart, yet, produced similar ideas and developed best practices of success. Both of us have started and/or acquired many companies, applied the entrepreneur philosophy, and made several successful exits because of our E-Myth structure and operating footprints. **This is a must-read for anyone searching for operational excellence in our ever-changing home service industry.**

Ben Stark, Go Time Success Group

Every contractor should read *The E-Myth HVAC Contractor*™ featuring ACCA member Ken Goodrich of Goettl Air Conditioning & Plumbing. Ken is an industry leader who went from the verge of losing his business, to being one of the fastest-growing HVAC contractors in the country. Ken has provided mentorship to countless peers in ACCA, served as a leader on ACCA's national board of directors, and continues to help fellow contractors develop their companies. I'm proud to know that Ken credits some of his success to being involved with ACCA, and especially the value of the Management Information Exchange (MIX) Group program. It is great to see Ken continuing to share his knowledge so posterity can learn from decades of experience and leadership in our industry. **I hope every contractor takes the opportunity to read *The E-Myth HVAC Contractor*™ and apply the business principles that helped Ken thrive.**

Barton James, President and CEO,
Air Conditioning Contractors of America (ACCA)

Ken Goodrich is a modern-day entrepreneur, a multidimensional businessman, and arguably the most recognized HVAC leader in the industry. He is a great salesman with a keen focus on the most granular details, giving him the unique ability to continually improve his business operations. I consider Ken's success in business to be highly correlated with his quest for answers and a willingness to listen to those around him.

Mark E. Berch, CEO, Service Finance Company, LLC

Being his competitor, I was afraid of the future of my business, knowing that a lifetime could not be enough to reach the level of knowledge required to compete with a proven performer like Ken Goodrich. I never imagined that such a successful person would share his secrets to improve someone else's business, even though it could mean (from an ordinary point of view) to help the competition to fight against him. Far from there, he got closer to me and became my personal mentor and taught me how to deliver results that I would never have reached on my own. **Now, from an extraordinary point of view, Ken is going even further by partnering with Michael E. Gerber making their knowledge available to any business person in the HVAC industry.** For this, I am deeply thankful, and better yet, I'm inspired to pay it forward.

Carlos Avemaria, President, Buenos Aires Air Conditioning

Michael E. Gerber's *The E-Myth*™ is one of only four books I recommend as required reading. **For those looking to start and build a business of their own, this is the man who has coached more successful entrepreneurs than the next ten gurus combined.**

Timothy Ferris, #1 *New York Times*
Best-selling author, *The 4-Hour Workweek*

Everyone needs a mentor, someone who tells it like it is, holds you accountable, and shows you your good, bad, and ugly. For millions of small business owners, Michael E. Gerber is that person. Let Michael be your mentor and you are in for a kick in the pants, the ride of a lifetime.

John Jantsch, author, *Duct Tape Marketing*

Michael E. Gerber is a master instructor and a leader's leader. As a combat F15 fighter pilot, I had to navigate complex missions with life-and-death consequences, but until I read *The E-Myth Revisited*™ and met Michael E. Gerber, my transition to the world of small business was a nightmare with no real flight plan. **The hands-on, practical magic of Michael's turnkey systems magnified by the raw power of his keen insight and wisdom have changed my life forever.**

Steve Olds, CEO, Patriot Mission

Michael E. Gerber's strategies in *The E-Myth Revisited*™ were instrumental in building my company from two employees to a global enterprise; I can't wait to see how applying the strategies from *Awakening the Entrepreneur Within* will affect its growth!

Dr. Ivan Misner, Founder and Chairman, BNI;
author, *Masters of Sales*

Michael E. Gerber's gift to isolate the issues and present simple, direct, business-changing solutions shines bright with *Awakening the Entrepreneur Within*™. **If you're interested in developing an entrepreneurial vision and plan that inspires others to action, buy this book, read it, and apply the processes Gerber brilliantly defines.**

Tim Templeton, author, *The Referral of a Lifetime*

Michael E. Gerber truly, truly understands what it takes to be a successful practicing entrepreneur and business owner. He has demonstrated to me over six years of working with him that for those who stay the course and learn much more than just "how to work on their business and not in it" then they will reap rich rewards. **I finally franchised my business, and the key to unlocking this kind of potential in any business is the teachings of Michael's work.**

Chris Owen, marketing director,
Royal Armouries (International) PLC

Michael's work has been an inspiration to us. **His books have helped us get free from the out-of-control life that we once had. His no-nonsense approach kept us focused on our ultimate aim rather than day-to-day stresses. He has helped take our business to levels we couldn't have imagined possible.** In the Dreaming Room™ made us totally re-evaluate how we thought about our business and our life. We have now redesigned our life so we can manifest the dreams we unearthed in Michael's Dreaming Room™.

Jo and Steve Davison, Founders, The Spinal Health Clinic
Chiropractic Group and www.your-dream-life.com

Michael E. Gerber is an outrageous revolutionary who is changing the way the world does business. **He dares you to commit to your grandest dreams and then shows you how to make the impossible a reality. If you let him, this man will change your life.**

Fiona Fallon, Founder, Divine and The Bottom Line

Michael E. Gerber is a genius. Every successful business person I meet has read Michael E. Gerber, refers to Michael E. Gerber, and lives by his words. You just can't get enough of Michael E. Gerber. **He has the innate (and rare) ability to tap into one's soul, look deeply, and tell you what you need to hear. And then, he inspires you, equips you with the tools to get it done.**

Pauline O'Malley, CEO, The RevTurbo Selling System
(formerly **'The Revenue Builder'**)

When asked "Who was the most influential person in your life?" I am one of the thousands who don't hesitate to say "Michael E. Gerber." **Michael helped transform me from someone dreaming of retirement to someone dreaming of working until age one hundred.** This awakening is the predictable outcome of anyone reading Michael's new book.

Thomas O. Bardeen

Michael E. Gerber is an incredible business philosopher, guru, perhaps even a seer. He has an amazing intuition, which allows him to see in an instant what everybody else is missing; he sees opportunity everywhere. **While in the Dreaming Room™, Michael gave me the gift of seeing through the eyes of an awakened entrepreneur, and instantly my business changed from a regional success to serving clients on four continents.**

Keith G. Schiehl, President, Rent-a-Geek Computer Services

Michael E. Gerber is among the very few who truly understand entrepreneurship and small business. While others talk about these topics in the form of theories, methodologies, processes, and so on, Michael goes to the heart of the issues. **Whenever Michael writes about entrepreneurship, soak it in as it is not only good for your business, but great for your soul.** His words will help you to keep your passion and balance while sailing through the uncertain sea of entrepreneurship.

Raymond Yeh, Co-Author, *The Art of Business*

Michael E. Gerber forced me to think big, think real, and gave me the support network to make it happen. A new wave of entrepreneurs is rising, much in thanks to his amazing efforts and very practical approach to doing business.

Christian Kessner, Founder, Higher Ground Retreats and Events

Michael's understanding of entrepreneurship and small business management has been a difference maker for countless businesses, including Infusion Software. **His insights into the entrepreneurial process of building a business are a must-read for every small business owner.** The vision, clarity, and leadership that came out of our Dreaming Room™ experience were just what our company needed to recognize our potential and motivate the whole company to achieve it.

Clate Mask, President & CEO,
Infusion Software

Michael E. Gerber is a truly remarkable man. His steady openness of mind and ability to get to the deeper level continues to be an inspiration and encouragement to me. **He seems to always ask that one question that forces the new perspective to break open and he approaches the new coming method in a fearless way.**

Rabbi Levi Cunin, Chabad of Malibu

The Dreaming Room™ experience was literally life-changing for us. **Within months, we were able to start our foundation and make several television appearances owing to his teachings.** He has an incredible charisma, which is priceless, but above all Michael E. Gerber awakens passion from within, enabling you to take action with dramatic results . . . starting today!

Shona and Shaun Carcary,
Trinity Property Investments Inc.
Home Vestors franchises

I thought E-Myth was an awkward name! What could this book do for me? **But when I finally got to reading it . . . it was what I was looking for all along.** Then, to top it off, I took a twenty-seven-hour trip to San Diego just to attend the Dreaming Room™, where Michael touched my heart, my mind, and my soul.

Helmi Natto, President,
Eye 2 Eye Optics, Saudi Arabia

I attended In the Dreaming Room™ and was challenged by Michael E. Gerber to "Go out and do what's impossible." So I did; **I became an author and international speaker and used Michael's principles to create a world- class company that will change and save lives all over the world.**

Dr. Don Kennedy, MBA; author, *5 AM & Already Behind*, www.bahbits.com

The  Myth

HVAC Contractor

*Why Most HVAC
Companies Don't Work
and What to Do About It*

MICHAEL E. GERBER
KEN GOODRICH

PRODIGY
BUSINESS BOOKS

Published by
Michael E. Gerber Partners™, Carlsbad, California.

Production Team
Helen Chang, Editor, AuthorBridgeMedia.com; Erich Broesel, cover designer, BroeselDesign, Inc.; Nancy Ratkiewich, book production, njr productions; Jeff Kassebaum, Michael E. Gerber author photographer, Jeff Kassebaum and Co.

For general information on other products and services, please visit the website: www.MichaelEGerberCompanies.com

ISBN 978-1-61835-040-4 (cloth)
ISBN 978-1-61835-041-1 (audio)
ISBN 978-1-61835-007-7 (ebook)

Printed in the United States of America

10 9 8 7 6 5 4 3 2 1

To Luz Delia, my partner, my wife, my inspiration, and my life . . .
Thank you for your perseverance, your indomitable will,
Your kind and generous Soul . . .
You're the Love of
my life . . . !

—Michael E. Gerber

With great pride, I dedicate this book to my father,
J. Duncan Goodrich. He worked hard to provide a good life for
our family and took care of his neighbors, customers, and friends.
He taught me the value of hard work, determination,
and a great respect for the HVAC trade. J. Duncan Goodrich
did things the right way, not the easy way.

—Ken Goodrich

CONTENTS

A WORD ABOUT THIS BOOK

Michael E. Gerber

I launched my business career in 1975, without a thought about what I was going to be doing for the rest of my life!

A friend of mine, indeed, my brother-in-law, had founded an advertising agency to serve small hi-tech companies in Silicon Valley, and one of his clients was having a problem converting the leads his advertising was creating for him. He asked me to help.

At that time in my life I had absolutely no interest in small business.

Indeed, I was on another journey of another kind, completely foreign to anything I'd ever done before.

I told my brother-in-law, "Why me, Ace? I don't know anything about small business."

"You know more than you think you do," he said. "Just come along for the ride. Let's see what happens."

So, I say the very same thing to you, Dear Reader. "Just come along for the ride. Let's see what happens."

My first E-Myth book was published in 1986.

It was the beginning of a long and enthralling exercise, called, by me, waking everyone up!

I named it, *The E-Myth: Why Most Small Businesses Don't Work and What to Do About It*™.

The term, E-Myth, stood for the Entrepreneurial Myth™. The fact that most small business owners aren't truly the entrepreneurs everyone thinks they are, but what I came to call, "technicians suffering from an Entrepreneurial Seizure™!"

Since that book, and the company I created to provide business development services to its many readers, millions have read *The E-Myth*™ and the books that followed it, starting out with *The E-Myth Revisited*™, and continuing with the more than 30 books I've authored which followed on its heels, along with the tens upon tens of thousands of small business owners who have participated in our E-Myth Mastery programs, but so, so many more.

Since *The E-Myth Revisited*™ took the business marketplace by storm, it became obvious to me that there was needed another series of books which addressed the application of my E-Myth Protocol™ in an entire subset of small companies, each of who have been attracted to *The E-Myth*™ and applied its thinking to the extreme development of their uniquely designed Practices. We came to call them Vertical Markets. Such as the Practice of Law, or Chiropractic, or Landscape Contracting, et al. There are literally thousands of them.

There are now eighteen of those "Vertical E-Myth books" addressing the seemingly unique problems of every field of endeavor, CoAuthored by those like Ken Goodrich of this book, who discovered early on that my E-Myth Protocol™ could be applied successfully, dramatically, and systematically, to each and every one of those Vertical Markets, to each and every industry on the planet.

So it is that our CoAuthors, like Ken Goodrich, have become the most avid fans of our work, having told the story of how it has worked to grow their companies to a degree never thought possible before.

That's when the idea for our next book series took form. Our C-Level Series. What I think of as the Core Operating Series™, CoAuthored by Functional experts in each of the most critical strategic functions needing to be filled in every small growing company.

I think of those Strategic Functions as Financial, Legal, Operational, Strategic Leadership, Marketing, Technological, Social, and Administrative.

Think Chief Financial Officer (CFO), Chief Operating Officer (COO), Chief Information Officer (CIO), Chief Technology Officer (CTO), Chief Marketing Officer (CMO), Chief Legal Officer (CLO), Chief Administrative Officer, (CAO).

In short, think getting one's house in order across the broad expanse of the Strategic Reach and Strategic Functionality of any company determined to grow beyond the flat level existence every Company of One lives at as it attempts to grow from one guy "doing it, doing it, doing it" to many guys and ladies struggling to figure out what is takes, what they've got to know to fill the shoes of the Growers at the top of their Company, let alone at the bottom, to turn it into a Great Growing Enterprise as each of our CoAuthors have so exquisitely done. From tiny to transformational. Just as Ken and I share with you in the pages that follow.

In short, think *"growth"*. Because, when you think "growth" you'll think about the essential functions you'll need in order to realize growth.

And when you think about the essential functions you'll need, you'll immediately realize that those functions are as critical to you today when you're small, as they will be in the future when you're not.

Indeed, as every single small business owner we've worked with over the past 40-plus years came to realize—yes, hundreds of thousands of them!—the great growing company they'll soon become hot on the trail of is a complete mystery to them there at the beginning of that intrepid journey.

And as that mystery becomes solved, for each and every one of them, a new life, a magical reality, a remarkable breakthrough comes sizzling toward them, changing absolutely everything, including all the people around them!

It is our position that to be inordinately successful, as Ken Goodrich so obviously has, every small company owner needs to assume the responsibility for each of these Eight Strategic Roles™, until such time as he or she is able to replace him or herself with the individuals designated to fill those roles.

Since there's no better way of accomplishing an objective than by simply getting started with it, let's get started.

Allow me to introduce you to Ken Goodrich, a brilliant entrepreneur, who is committed to E-Myth Systems™ and creating profitable HVAC businesses.

Welcome to *The E-Myth HVAC Contractor*™!

What a brilliant opportunity for each of us to grow exponentially and to understand how that works.

—Michael E. Gerber
CoFounder | Chairman | Chief Dreamer
Michael E. Gerber Companies, Inc.™
Carlsbad, California
MichaelEGerberPartners.com

A NOTE FROM KEN

Ken Goodrich

Heating, ventilation, and air conditioning (HVAC) is not a luxury. It's a life support system. As an HVAC contractor, you hold the power of life and death in your hands. In extremely hot weather, people can die of heatstroke. In extremely cold weather, they can die of exposure.

As an HVAC contractor, you have a noble profession. You are part of an honorable industry. You make life possible in certain parts of the world; in others, you save lives.

As an entrepreneur, you can live with honor, respect, and prosperity. You have the opportunity to run a successful business, have time for your family, and live a happy, accomplished, abundant life.

Maybe those were your dreams when you started out, but somehow, it hasn't turned out that way. Maybe you feel stuck, overwhelmed, or frustrated by the business. You're working early hours and late nights, and you feel trapped in the business. Maybe you feel financially burdened. You've got customers, but as soon as money comes in one hand, it goes out the other, and you can't seem to get ahead.

Maybe you feel disappointed by the pace of your business's growth. You seem unable to hire the right people or keep them when you do. Maybe you've even hired people with impressive résumés, but no results. They just didn't understand your business.

Or maybe you're doing well, but you can't seem to expand to that next level of growth you always dreamed of. You don't know how.

Managing people, money, and time seems overwhelming. Or maybe you just want to succeed faster. You're doing five, six, or seven figures in business every year, but you want to grow to the next goal. And actually make a profit.

Or maybe you just want more in life. You want more time for your spouse and kids. You want to take them on vacations. You want to be able to relax, without the business calling you. You want to save for your future.

I get it. Wherever you are in your business, I've probably been there. When I started as a business owner, I became a master at fixing air conditioners and installing heaters, but I didn't know how to run a business. I thought my HVAC skills translated into business skills. But I was wrong.

Learning about Business

I started in the HVAC business at age ten, working alongside my father. I bought the family company from my mother at age twenty-five, after my father passed.

But I didn't really learn to run the business until I read Michael E. Gerber's *The E-Myth Revisited*™ at age twenty-seven. By then, I was stonewalled by vendors, shunned by suppliers, deserted by employees, and $75,000 in debt to the IRS. I was also cocky, resentful, and ignorant. A lifelong friend had given me *The E-Myth Revisited*™, but the book sat on my shelf for three months. Finally, the pain from my business was so bad, I opened it.

Its pages astonished me. *How does Gerber know exactly what I'm going through?* I thought. *He must have been spying on me to write this book!* I read the entire book in one night. As I pored over each page, a sense of peace and understanding flowed over me. Maybe I was not so different from other business owners. Maybe the mistakes I had made were not uncommon.

By the time I finished reading the book, I had realized that it contained the solutions to all my problems. I kept a copy of the book in my back pocket at all times. It became my trusted guide. When a vendor shamed me, I pulled out the book and wrote down plans on my yellow notepad. When employees quit, I read its pages and created new systems.

Every day, I looked myself in the mirror and vowed to end the agony of just "doing it," over and over, every day, and instead accomplish something by creating an E-Myth prototype HVAC business.

I learned to manage systems that empower people. I learned to price correctly, manage cash flow, collect on a daily basis, and create profits. I learned to stop thinking like a technician and manager and begin thinking like an entrepreneur. It had taken me years to become a master HVAC contractor, so I knew it would take me more time to become a master entrepreneur.

In three years, I paid off the IRS debt, which had grown to $300,000. I stabilized my business and soon started to lead the HVAC market in my area.

Dreams and Plans

More importantly, I allowed myself to dream and plan. When I first read *The E-Myth*™ book, I wrote down a goal to sell my business in five years for $1 million. Five years later, I sold not just one, but three HVAC businesses—for more than six times my goal.

After I took a couple of years off, my team and I began building and monetizing HVAC businesses throughout the Southwest USA, which by 2019 included twenty-four HVAC and plumbing operations.

During this time, I acquired the legacy company that had manufactured the very first air conditioner I had ever worked on and the one my dad's company specialized in: Goettl Air Conditioning. I believe my collective experiences in business have led me to this destiny, to perpetuate the Goettl brand and honor its founders and thousands of team members who came before and after me to build this enduring great company.

As of this writing, the Goettl Home Services group of companies does more than $100 million in sales each year, and we're focused on becoming a national service provider. Our mission: Goettlize the Nation™.

By now, I've read Michael E. Gerber's *The E-Myth Revisited*™ thirty-nine times, and I've given away more than one hundred copies. This book, *The E-Myth HVAC Contractor*™, is my way to repay my mentor, Michael E. Gerber, of what he so miraculously provided me decades ago. It's also a way for me to pay it forward, to give you and the industry what I learned, so you can avoid my mistakes and catapult your growth.

Now, you have two people who secretly know your business—Michael E. Gerber and me. You can be a master HVAC contractor and a master entrepreneur.

You hold the power of life and death in your hands—both as an HVAC contractor and as an entrepreneur. With this book, *The E-Myth HVAC Contractor*™, you have the power to give your business new life while giving life to others.

—Ken Goodrich
Chief Executive Officer
Goettl Home Services
Las Vegas, Nevada

PREFACE

Michael E. Gerber

O ver the past four decades, it has been my delight to inspire small business owners in every trade and profession to acquire and apply the methods and skills we've invented and perfected for the purpose of growing and thriving in their role as entrepreneur.

Most important in our work has been the Dream, Vision, Purpose, and Mission that resides at the heart of it, without which nothing we've accomplished for the millions of small business owners we've reached out to and touched would have, could have, made any difference.

So too has that unadulterated passion been the source for our new, online Entrepreneurial Development School, Radical U™, whose purpose it is to transform the state of entrepreneurship and small business development in a manner never achieved or ever attempted, now available worldwide.

As you, an HVAC contractor, read Ken Goodrich's story, and the absolutely profound success he's created utilizing my E-Myth Philosophy, Paradigm and tools, know that at the foundation of it all is the entrepreneurial energy and imagination now available to you all at Radical U™.

Ken and I invite you to join us by simply sharing your experiences and insights with us, so we can all grow as an industry and society. Join our email list, attend our workshops, or hear me speak at www.TheEMythHVAC.com. Welcome to the extraordinary world of INSPIRATION.

Welcome to the first day of the rest of your life!

—Michael E. Gerber
CoFounder | Chairman | Chief Dreamer
—Luz Delia Gerber
CoFounder | CEO | President
Michael E. Gerber Companies, Inc.™
Radical U™
The New Dreaming Room™
Carlsbad, California

ACKNOWLEDGEMENTS

Michael E. Gerber

A s always, and never to be forgotten, there are those who give of themselves to make my work possible.

To my dearest and most forgiving partner, wife, friend, and CoFounder, Luz Delia Gerber, whose love and commitment takes me to places I would often not go unaccompanied.

To Patricia Beaulieu, whose continual commitment to The GerberWorks™ is beyond Words. Thank you Trish!

To Helen Chang, noble warrior, editor, brave soul, and sojourner, who covers all the bases we would have missed had she not been there.

And to Nancy Ratkiewich, whose work has been essential for you who are reading this.

To those many, many Dreamers, Thinkers, Storytellers, and Leaders, whose travels with me in The Dreaming Room™ have given me life, breath, and pleasure unanticipated before we met. To those many participants in my life (you know who you are), thank you for taking me seriously, and joining me in this exhilarating quest.

And, of course, to my CoAuthors, all of you, your genius, wisdom, intelligence, and wit have supplied me with a grand view of the world, which would never have been the same without you.

Love to all.

Michael

ACKNOWLEDGMENTS

Ken Goodrich

To my beautiful wife and children, Wendy, Megan, and Duncan: Thank you for giving me the love, time, support, encouragement, poking, and prodding that allowed me to build our businesses. We are the dream team!

To Ben DiIorio: a true friend and consigliere. From the beginning, you were always there for us no matter what challenges we faced.

To G. Cash Wilson: my lifelong friend, with genuine intentions you gave me the gift of E-Myth knowledge.

To my brothers of Kappa Sigma, Kappa Alpha Chapter, AEKDB.

To Roy H. Williams, Dave Slott, Richard Haddrill, and Dan Burke: I thank each one of you for freely imparting your experience and knowledge to me at critical points in my journey. The gift of your precious time, guidance, and mentorship has helped me achieve well beyond what I ever thought possible. I will continue to be a good steward of your teachings and will most certainly pay it forward to my peers.

To my fellow members of serviceroundtable.com and Service Nation Alliance: by embracing the vision and technology of Matt Michel and David Heimer, we have openly shared, accelerated our performance, supported each other, and thereby made each other better men and women, business people and industry leaders.

To my key people and partners over the years: Mike Luzader, Curt Coker, Jon Catalano, Lance Fernandez, Bill Moore, Jeremy Prevost, Allen Crick, Colin Martodam, Dan Traversi, Landon Brewer, Angela

Miller, Darin Wetzler, Dale Steele, Stephen Gamst, Clayton Johnson, Rohin Lal, and Karl Pomeroy: No matter what road each of us is on today, I will always fondly remember the world-class teams we built and the astounding results we created together.

To the members and staff of the Air Conditioning Contractors of America, (ACCA). From the first day in the HVAC contracting business, my team has utilized the industry defining tools and business systems that ACCA created and made available to its members. Those tools and business systems differentiated my companies from the competition and made us known as experts in the business and science of HVAC. More importantly, the friendship and collaboration with my fellow ACCA members proved to be a valuable resource and one of the most important keys to our success.

To Helen Chang and the Author Bridge Media team: Your commitment to process in your business, just as the E-Myth describes, has made this experience fun, thought-provoking, and exhilarating! I can't wait to partner with you on my next books!

From the bottom of my heart, thank you, Michael E. Gerber, for your brilliant insight and invaluable mentorship.

INTRODUCTION

Michael E. Gerber

As I write this book, the marketplace continues to do what it has always done: it goes up and it goes down. Today, in 2019, it's feverishly bullish.

Tomorrow, it could tear your heart out.

What has always been true, however, is that the marketplace lives by rules great companies don't. The marketplace is an emotional storm, where rules one day defy attitudes the very next. Where the optimistic truth on Monday is beset upon by the terrorized truth like a ferocious dog on Tuesday.

If one were to run a company like the marketplace runs, disaster would be the common experience.

Come to think of it, while great companies do exactly the *opposite* of what the marketplace does, most companies unfortunately aren't great companies.

And that's why the number of small companies that start up in any one year almost equals the number of small companies that close their doors.

Think about it: for every one hundred new companies started, half of them fail in that very same year!

And it doesn't matter what kind of company it is, either.

Which means that when a company, small or large, operates according to emotional decisions, as all struggling, disorganized, dismally managed companies do, that company is doomed to fail.

And from the very beginning to the very end, the vast majority of small companies are purely operated by emotional decisions, choices, preferences, opinions, attitudes, beliefs, and hopes. Thus, this book, and Ken Goodrich.

Ken Goodrich is unlike anyone I've ever met.

Truly, an original.

His HVAC company has been designed, built, launched, and grown, word for word, idea for idea, inspiration for inspiration, system for system, based completely on his interpretation and the inspiration driven into him by *The E-Myth Revisited*™, so influentially that, as Ken says it, "I carried it in my pocket everywhere I went. I read it thirty-nine times!"

Can you believe that? Well, do believe that! Because Ken only tells the truth. He's that kind of guy. Truly, an original!

And he did the E-Myth so ferociously, so adamantly, so devotedly, that he wore that book out, dozens upon dozens of times! Remarkable, wouldn't you say? Well, even more remarkable is that, Ken turned his first absolute failure as an HVAC contractor into a $100 million annual raving success!

This is something you can do, something everyone can do, something we've done with *The E-Myth*™ again and again and again in every industry.

But, here's my introduction to you. Do what Ken did. Do it with the passion Ken infused into every step he took, from his Company of One to his Company of 1,000! Read every single chapter in this book, and then join us in Radical U™.

Your life is about to change, dramatically. Just like Ken's did. Just like Ken's is continuing to!

Now, let Ken and I help you, every single step of the way forward.

To your ever-improving life, and business. Welcome to *The E-Myth HVAC Contractor*™.

1

The Story of Steve and Peggy

Michael E. Gerber

Make every detail perfect, and limit the number of details to perfect.
—Jack Dorsey, CoFounder of Twitter

E very business is a family business. To ignore this truth is to court disaster.

I don't care if family members actually work in the business or not. Whatever his or her relationship with the business, every member of a HVAC contractor's family will be greatly affected by the decisions the contractor makes about the business. There's just no way around it.

Unfortunately, like most contractors, HVAC contractors tend to compartmentalize their lives. They view their trade as a profession—what they do—and therefore none of their family's business.

"This has nothing to do with you," says the contractor to his wife, with blind conviction. "I leave work at the office and family at home."

And with equal conviction, I say, "Not true!"

1

In actuality, your family and contracting business are inextricably linked to one another. What's happening in your business is also happening at home. Consider the following and ask yourself if each is true:

If you're angry at work, you're also angry at home.

If you're out of control at your contracting business, you're equally out of control at home.

If you're having trouble with money at your company, you're also having trouble with money at home.

If you have communication problems at your company, you're also having communication problems at home.

If you don't trust in your company, you don't trust at home.

If you're secretive at your company, you're equally secretive at home.

And you're paying a huge price for it!

The truth is that your business and your family are one—and you're the link. Or you should be. Because if you try to keep your business and your family apart, if your business and your family are strangers, you will effectively create two separate worlds that can never wholeheartedly serve each other. Two worlds that split each other apart.

Let me tell you the story of Steve and Peggy Walsh.

The Walshes met in trade school. They were lab partners in basic electrical training, Steve an HVAC student and Peggy a landscape technician student. When their lab discussions started to wander beyond safety and circuitry into their personal lives, they discovered they had a lot in common. By the end of the course, they weren't just talking in class; they were talking on the phone every night ... and *not* about transformers and fuses.

Steve thought Peggy was absolutely brilliant (her love for dogs was astounding), and Peggy considered Steve the most passionate man she knew. It wasn't long before they were engaged and planning their future together. A week after graduation, they were married in a lovely garden ceremony in Peggy's childhood home.

While Steve apprenticed with a nearby HVAC installation company, Peggy entered an ornamental horticulture program nearby.

Over the next few years, the couple worked hard to keep their finances afloat. They worked long hours and studied constantly; they were often exhausted and struggled to make ends meet. But throughout it all, they were committed to what they were doing and to each other.

After passing the EPA section 608 exam, Steve went on to get a specialized degree in air distribution while Peggy began working at a large landscape contracting company nearby. Then Steve started working for a large, publicly traded commercial HVAC company. Soon afterward, the couple had their first son, and Peggy decided to take some time off from the field to be with him. Those were good years. Steve and Peggy loved each other very much, were active members in their church, participated in community organizations, and spent quality time together. The Walshes considered themselves one of the most fortunate families they knew.

But work became troublesome. Steve grew increasingly frustrated with the way the company was run. "I want to go into business for myself," he announced one night at the dinner table. "I want to start my own company."

Steve and Peggy spent many nights talking about the move. Was it something they could afford? Did Steve really have the skills necessary to make an HVAC contracting business a success? Were there enough customers to go around? What impact would such a move have on Peggy's career at the landscape company, their lifestyle, their son, their relationship? They asked all the questions they thought they needed to answer before Steve went into business for himself . . . but they never really drew up a concrete plan.

Finally, tired of talking and confident that he could handle whatever he might face, Steve committed to starting his own HVAC contracting business. Because she loved and supported him, Peggy agreed, offering her own commitment to help in any way she could. So Steve quit his job, took out a second mortgage on their home, and leased a small office nearby.

In the beginning, things went well. A housing boom had hit the town, and new families were pouring into the area. Steve had no trouble getting new customers. His business expanded, quickly outgrowing his office.

Within a year, Steve had employed an office manager, Clarissa, to book appointments and handle the administrative side of the business. He also hired a bookkeeper, Tim, to handle the finances. Steve was ecstatic with the progress his young business had made. He celebrated by taking his wife and son on vacation to Italy.

Of course, managing a business was more complicated and time-consuming than working for someone else. Steve not only supervised all the jobs Clarissa and Tim did, he was continually looking for work to keep everyone busy. When he wasn't scanning trade journals to stay abreast of what was going on in the field or fulfilling continuing-education requirements to stay current on his certifications, he was going to the bank, wading through customer paperwork, or speaking with vendors (which usually degenerated into *arguing* with vendors). He also found himself spending more and more time on the telephone dealing with customer complaints and nurturing relationships.

As the months went by and more and more customers came through the door, Steve had to spend even more time just trying to keep his head above water.

By the end of its second year, the company, now employing two full-time and two part-time people, had moved to a larger office downtown. The demands on Steve's time had grown with the company.

He began leaving home earlier in the morning, returning home later at night. He drank more. He rarely saw his son anymore. For the most part, Steve was resigned to the problem. He saw the hard work as essential to building the "sweat equity" he had long heard about.

Money was also becoming a problem for Steve. Although the business was growing like crazy, money always seemed scarce when it was really needed.

When Steve had worked for somebody else, he had been paid twice a month. In his own business, he often had to wait—sometimes for months. He was still owed money on billings he had completed more than ninety days before.

Of course, no matter how slowly Steve got paid, he still had to pay *his* people. This became a relentless problem. Steve often felt like a juggler dancing on a tightrope. A fire burned in his stomach day and night.

To make matters worse, Steve began to feel that Peggy was insensitive to his troubles. Not that he often talked to his wife about the company. "Business is business" was Steve's mantra. "It's my responsibility to handle things at the office and Peggy's responsibility to take care of her own job and the family."

Peggy herself was working weekend hours at the landscape company, and they'd brought in a nanny to help with their son. Steve couldn't help but notice that his wife seemed resentful, and her apparent lack of understanding baffled him. Didn't she see that he had a company to take care of? That he was doing it all for his family? Apparently not.

As time went on, Steve became even more consumed and frustrated by his business. When he went off on his own, he remembered saying, "I don't like people telling me what to do." But people were still telling him what to do.

Not surprisingly, Peggy grew more frustrated by her husband's lack of communication. She cut back on her own hours to focus on their family, but her husband still never seemed to be around. Their relationship grew tense and strained. The rare moments they were together were more often than not peppered by long silences—a far cry from the heartfelt conversations that had characterized their relationship's early days, when they'd talk into the wee hours of the morning.

Meanwhile, Tim, the bookkeeper, was also becoming a problem for Steve. Tim never seemed to have the financial information Steve needed to make decisions about payroll, customer billing, and general operating expenses, let alone how much money was available for Steve and Peggy's living expenses.

When questioned, Tim would shift his gaze to his feet and say, "Listen, Steve, I've got a lot more to do around here than you can imagine. It'll take a little more time. Just don't press me, okay?"

Overwhelmed by his own work, Steve usually backed off. The last thing Steve wanted was to upset Tim and have to do the books himself. He could also empathize with what Tim was going through, given the business's growth over the past year.

Late at night in his office, Steve would sometimes recall his first years out of school. He missed the simple life he and his family had shared. Then, as quickly as the thoughts came, they would vanish. He had work to do and no time for daydreaming. "Having my own company is a great thing," he would remind himself. "I simply have to apply myself, as I did in school, and get on with the job. I have to work as hard as I always have when something needed to get done."

Steve began to live most of his life inside his head. He began to distrust his people. They never seemed to work hard enough or to care about his company as much as he did. If he wanted to get something done, he usually had to do it himself.

Then one day, the office manager, Clarissa, quit in a huff, frustrated by the amount of work that her boss was demanding of her. Steve was left with a desk full of papers and a telephone that wouldn't stop ringing.

Clueless about the work Clarissa had done, Steve was overwhelmed by having to pick up the pieces of a job he didn't understand. His world turned upside down. He felt like a stranger in his own company.

Why had he been such a fool? Why hadn't he taken the time to learn what Clarissa did in the office? Why had he waited until now?

Ever the trouper, Steve plowed into Clarissa's job with everything he could muster. What he found shocked him. Clarissa's work space was a disaster area! Her desk drawers were a jumble of papers, coins, pens, pencils, rubber bands, envelopes, business cards, contact lenses, eye drops, and candy.

"What was she thinking?" Steve raged.

When he got home that night, even later than usual, he got into a shouting match with Peggy. He settled it by storming out of the house to get a drink. Didn't anybody understand him? Didn't anybody care what he was going through?

He returned home only when he was sure Peggy was asleep. He slept on the couch and left early in the morning, before anyone was awake. He was in no mood for questions or arguments.

When Steve got to his office the next morning, he immediately headed for the makeshift kitchen, nervously looking for some Tylenol to get rid of his throbbing headache.

What lessons can we draw from Steve and Peggy's story? I've said it once and I'll say it again: *Every business is a family business.* Your business profoundly touches every member of your family, even if they never set foot inside your office. Every business either gives to the family or takes from the family, just as individual family members do.

If the business takes more than it gives, the family is always the first to pay the price.

In order for Steve to free himself from the prison he created, he would first have to admit his vulnerability. He would have to confess to himself and his family that he really didn't know enough about his own business and how to grow it.

Steve tried to do it all himself. Had he succeeded, had the company supported his family in the style he imagined, he would have burst with pride. Instead, Steve unwittingly isolated himself, thereby achieving the exact opposite of what he sought.

He destroyed his life—and his family's life along with it.

Repeat after me: Every business is a family business.

Are you like Steve? I believe that all contractors share a common soul with him. You must learn that a business is only a business. It is not your life. But it is also true that your business can have a profoundly negative impact on your life unless you learn how to do it differently than most contractors do it—and definitely differently than Steve did it.

Steve's HVAC business could have served his and his family's life. But for that to happen, he would have had to learn how to master his business in a way that was completely foreign to him.

Instead, Steve's business consumed him. Because he lacked a true understanding of the essential strategic thinking that would have allowed him to create something unique, Steve and his family were doomed from day one.

This book contains the secrets that Steve should have known. If you follow in Steve's footsteps, prepare to have your life and business fall apart. But if you apply the principles we'll discuss here, you can avoid a similar fate.

Let's start with the subject of *money*. But, before we do, let's read an HVAC contractor's view about the story I just told you. Let's talk about Ken's journey . . . and yours. ❧

The Boy with the Flashlight

Ken Goodrich

Money for nothin' and chicks for free, (I want my, I want my MTV).
—Dire Straits, *Money for Nothing*

No matter where you are in your business journey, you are not alone. Countless entrepreneurs and HVAC contractors have probably been there, too. At the same time, countless entrepreneurs and HVAC contractors, such as I, have succeeded in using Michael E. Gerber's principles.

With my first company in 1988, I struggled with payroll and was $300,000 in debt. After applying Michael's principles, I turned that company around and, over the next thirty years, monetized twenty-four HVAC and plumbing companies.

By 2019, I owned a group of companies doing more than $100 million in sales. We employed more than four hundred team members, operated in three states, and had more than 250 service and installation trucks in our fleet.

Michael E. Gerber and *The E-Myth Revisited*™ principles made my success possible. In reading my story, I hope you can relate to my experiences and find inspiration in my growth. Here's my journey.

At First Light

I began my training as an HVAC technician when I was ten years old.

"Hey, I need you to help me," my dad told me one evening. "We gotta work on the neighbor's air conditioner."

My dad was an engineer by day and an air conditioner repairman by night. With his growing business, he needed my help.

At the neighbor's house, I took on the task my dad most needed help with that night: holding the flashlight. That's how I became the Official Flashlight Holder, a job I was to hold for years to come, as I followed in my dad's footsteps in the business.

We lived in Las Vegas, where the desert heat can kill you. Most air conditioners in the 1970s didn't work in the high heat of the Southwest. Only one brand was designed to work in the desert: Goettl. My dad loved Goettl air conditioners. That was his preferred brand to sell and install for his customers.

The first air conditioner I ever shined my flashlight on was a Goettl. I grew up working on, installing, and selling Goettl air conditioners. That became my brand, too. When I grew up, it was my brand for the majority of the companies I owned and sold until Goettl stopped building them in 2007.

My Dad's Company

In 1977, when I was fifteen years old, my dad retired from his day job to run the HVAC business full time. Over the years, dad repaired things like cars and televisions, but with Las Vegas booming, he found the greatest opportunities in air conditioning. We worked evenings, weekends, and every summer to serve customers.

My mom became the business manager and co-owner. She answered phones, made us lunch, and served us hot dinners, no matter how late we finally got home. She also tracked the inventory and made sure to take care of the books.

I never wanted to be an HVAC contractor, but I enjoyed pitching in to help my mom and dad in the business. Dad, the perfectionist HVAC master, made sure that I became a great technician. Together, we tackled the jobs that nobody else could handle. As soon as I got a driver's license at age sixteen, I was sent in my own van to make service calls. I made sales, completed service calls, installed air conditioners, and even built the sheet metal ductwork.

I worked so much that I envied my high school friends. During summer vacations, they were jumping in the pool playing Marco Polo, while I was on the roof fixing their air conditioners. As I watched them one day, sweat pouring down my brow, I decided that being an HVAC contractor was definitely not my dream job. I wanted to go to college.

MTV Dreams

Ever since I was a kid, I wanted to be a successful entrepreneur. When MTV came out, I imagined a lifestyle that looked like a 1980s MTV music video: loud music, poolside parties, and a mullet haircut. Yeah, that was my goal.

In college, I decided that a business degree would be the fastest way to reach that dream. While I studied finance on campus, I supported myself by fixing air conditioners off-campus. It wasn't the MTV lifestyle, but I was the most comfortable college student I knew.

As a finance major, I got job offers from big Wall Street investment firms. But to my surprise, what they paid in full-time salary was a pittance compared with what I made fixing air conditioners *on the side*. Working part-time, I was already making triple what starting bankers made. I realized that my own little business of selling, installing, and repairing air conditioners gave me the best opportunity to achieve my dreams. After all, I was an excellent technician. I could fix practically anything. I knew how to solve problems that most people couldn't or wouldn't solve.

Rather than pursue a career in the corporate world, I decided to build my own air conditioning company. But I could still have some semblance of the MTV lifestyle. Shortly after graduation, I got a mullet haircut.

On My Own

I was twenty-five years old when my father passed.

Dad never became wealthy, but he called himself "comfortable." While I loved and respected my dad, I believe that "comfortable" came at a high cost. He worked eighty hours a week, and it cost his health, peace of mind, and happiness. At the time, I had just started my own HVAC company. As I mourned the loss, I decided that my company was going to be different, not just a job where I made a good living.

Tasting the MTV Lifestyle

Even though I had made money as an HVAC technician on the side, I didn't really know how to run the business full time.

When I started my business, Goettl was the only company that gave me credit. It was the only place I could get parts without paying in advance. I like to think it was the mullet haircut that got me the credit.

Within two years, my business had grown to twenty-four employees and a fleet of new vans with fancy lettering on them. Our company signs went up all over Las Vegas.

I was young and energetic, and so was my team. We were aggressive with our marketing in an era when other HVAC businesses were not. We had big signs all over town with our name on many construction projects. Our new white vans stood out on the city streets with fancy lettering.

"Wow," people exclaimed. "I see you guys everywhere!"

With all my success, I decided it was time to get out of a service van and into a new BMW. After all, I was the business owner. I dressed in fancy new suits and ties while driving my fancy new BMW. I took two-hour lunches, hob-nobbing with other important people around town.

Every day, I had a sense of excitement. *I'm living the dream! I'm the master of the HVAC world! I can taste the MTV lifestyle!* One of the people I lunched with was my best friend from high school, Cash Wilson, who was working to build his father's architecture business. I had often chatted with him about plans to buy the family business, and now we continued to chat about my growing business.

I had learned a lot of business theory and ideas in college, but running an actual business was very different. I joked about it between bites, but the truth was that even though my company was known all over town, I struggled every week to make payroll. My employees always seemed to want too much, and customers complained about stupid stuff.

I was really good at selling and installing air conditioners, but I was overwhelmed and frustrated by running the business. I didn't want to work eighty-hour weeks as my father had, but I didn't know how else to grow. My friend seemed like one of the few people I could confide in.

One day, Cash handed me a book. "Before you run one more service call," he said, "I suggest you read this book—three times."

When I got back to the office, I put the book on my fancy new office shelf, along with all the other books I never read. The book was *The E-Myth Revisited*™, by Michael E. Gerber.

The Empty Suit

One day, a disheveled man in his fifties pulled up at my office. He drove a beat-up car. I didn't think much of him. He lacked my energy, style, and fancy BMW.

He introduced himself curtly and handed me a business card. It read: "Internal Revenue Service."

I looked up at him as he spoke. "You haven't been paying your payroll taxes," he said.

I stared back at him. "What's payroll taxes?"

I relied on my bookkeeper to handle those kinds of things. Later, I confronted her in the back office.

"Why didn't you pay the payroll taxes?" I yelled.

"Because we never have the money to pay it!" she yelled back.

When the IRS agent went through our shoddy accounting records, he demanded that I agree to a payment plan.

"We will not accept less than $10,000 a month," he said.

It seemed like a lot of money, but I didn't know how much money I was actually making or how much I could really afford. It never occurred to me to seek business or legal advice.

"Okay," I said.

I was two days late making the first IRS payment. *Big deal. Who's gonna notice?* I thought.

The next day, six agents with guns and badges arrived at my office. They said they were there to seize my property for not paying the IRS on time.

Outside, they slapped big orange stickers that said, "IRS SEIZURE!" all over our six new vans with fancy lettering. Then they hooked them up to tow trucks and drove them away.

The IRS also seized all the money in my bank account. Our employee paychecks and vendor checks all bounced.

My top technicians confronted me in the office. "We're not working until we get paid," they said.

"We have the work on the schedule to get you paid!" I pleaded. "Just get back out there and finish it, and I can pay you on Monday!"

Most of my employees quit on the spot. Soon, vendors showed up, demanding payment.

When the day finally ended, I crawled upstairs to my fancy office, sat in my fancy chair, and begged for relief somehow. I was ashamed and defeated, but too afraid to give up. I had promised my mother that I would pay for my father's assets. I had my father's reputation of twenty-five years in the business to live up to. And I was determined to prove the naysayers wrong. I held my head in my hands.

When I finally looked up again, I noticed the book my friend had given me three months before. I pulled it off the shelf and opened its pages.

Michael spoke to my heart. Miraculously, he knew everything about my business. I earmarked page after page, soaking in sentence after sentence.

When I was done, I had a new vision for my business. I knew I could lead it to success. My business would be based on real business principles, as taught by Michael, not just fancy symbols of success. I would no longer be an empty suit.

Carrying Michael's book in my back pocket everywhere I went, I would lead the business back to profitability, honor, and respect.

Full Circle

Three years later, I had paid back the IRS in full. With penalties and interest, the total had ballooned into nearly $300,000. We also became the market leader in the residential service and replacement arena in Las Vegas. Two years after that, I sold my three companies for six times my original goal.

Over the next thirteen years, I embarked on an incredible journey of acquiring underperforming HVAC companies, increasing their profitability, and monetizing them by combining my eclectic experiences of what *not* to do with Michael E. Gerber's teachings of what *to* do.

In 2013, I bought an ailing air conditioning business. It was doing $11 million in sales but losing $3 million a year. That business was Goettl Air Conditioning, the beloved air conditioning company of my childhood. Six years after the purchase, using the very same principles Michael E. Gerber taught, I led the company to $100 million in sales with a 17 percent profit. I had come full circle.

By the time I sold my first companies, I could easily have had the MTV lifestyle I had always dreamed of. But my mullet was gone, and

I had learned that success wasn't about poolside parties or fancy suits. My adversity had taught me that success is measured by how many people's lives one can positively influence.

Over the years, with *The E-Myth Revisited*™ by my side, I grew from technician and manager to entrepreneur. I learned to build companies and then enterprises. I learned that a strong business is managed by systems that are carried out by a team. If I had replicable systems and a team that could execute, the business would become extremely valuable.

As Michael says, "The system is the solution."

In the chapters ahead, I show you the exact systems I created to build my businesses. These areas cover money, team, time, employees, change, and even action. They're all systems.

My lessons started with learning to manage money. First, let's hear Michael's insights about *money*. ✤

CHAPTER 3

On the Subject
of Money

Michael E. Gerber

*If money is your hope for independence, you will never have it. The only
real security that a man will have in this world is a reserve of knowledge,
experience, and ability.*

—Henry Ford

Had Steve and Peggy first considered the subject of *money* as
we will here, their lives could have been radically different.

Money is on the tip of every contractor's tongue, on the
edge (or at the very center) of every contractor's thoughts, intruding
on every part of a contractor's life.

With money consuming so much energy, why do so few contractors handle it well? Why was Steve, like so many contractors, willing
to entrust his financial affairs to a relative stranger? Why is
money scarce for most contractors? Why is there less money than
expected? And yet the demand for money is always greater
than anticipated.

What is it about money that is so elusive, so complicated, so difficult to control? Why is it that every contractor I've ever met hates to deal with the subject of money? Why are they almost always too late in facing money problems? And why are they constantly obsessed with the desire for more of it?

Money—you can't live with it and you can't live without it. But you'd better understand it and get your people to understand it. Because until you do, money problems will eat your business for lunch.

You don't need an accountant or financial planner to do this. You simply need to prod your people to relate to money very personally. From technician to receptionist, they should all understand the financial impact of what they do every day in relationship to the profit and loss of the company.

And so you must teach your people to think like owners, not like technicians or office managers or receptionists. You must teach them to operate like personal profit centers, with a sense of how their work fits in with the company as a whole.

You must involve everyone in the business with the topic of money—how it works, where it goes, how much is left, and how much everybody gets at the end of the day. You also must teach them about the four kinds of money created by the business.

The Four Kinds of Money

In the context of owning, operating, developing, and exiting from an HVAC business, money can be split into four distinct but highly integrated categories:

Income

Profit

Flow

Equity

Failure to distinguish how the four kinds of money play out in your business is a surefire recipe for disaster.

Important Notes: Do not talk to your accountants or bookkeepers about what follows; it will only confuse you. The information comes from the real-life experiences of thousands of small business owners, contractors included, most of whom were hopelessly confused about money when I met them. Once they understood and accepted the following principles, they developed a clarity about money that could only be called enlightened.

The First Kind of Money: Income

Income is the money a company pays its employees for doing their job *in* the business, including the HVAC contractor/owner. It's what they get paid for going to work every day.

Clearly, if HVAC contractors didn't do their job, others would have to, and *they* would be paid the money the business currently pays the contractors. Income, then, has nothing to do with *ownership*. Income is solely the province of *employeeship*.

That's why to the HVAC technician-as-*employee*, Income is the most important form money can take. To the HVAC contractor-as-*owner*, however, it is the least important form money can take.

Most important; least important. Do you see the conflict? The conflict between the technician-as-employee and the contractor-as-owner?

We'll deal with this conflict later. For now, just know that it is potentially the most paralyzing conflict in a contractor's life.

Failing to resolve this conflict will cripple you. Resolving it will set you free.

The Second Kind of Money: Profit

Profit is what's left over after an HVAC contractor's business has done its job effectively and efficiently. If there is no profit, the business is doing something wrong.

However, just because the business shows a profit does not mean it is necessarily doing all the right things in the right way. Instead, it just means that something was done right during or preceding the period in which the profit was earned.

The important issue here is whether the profit was intentional or accidental. If it happened by accident (which most profit does), don't take credit for it. You'll live to regret your impertinence.

If it happened intentionally, take all the credit you want. You've earned it. Because profit created intentionally, rather than by accident, is replicable—again and again. And your business's ability to repeat its performance is the most critical ability it can have.

As you'll soon see, the value of money is a function of your business's ability to produce it in predictable amounts at an above-average return on investment.

Profit can be understood only in the context of your business purpose, as opposed to *your* purpose as an HVAC contractor. Profit, then, fuels the forward motion of the business that produces it. This is accomplished in four ways:

Profit is *investment capital* that feeds and supports growth.

Profit is *bonus capital* that rewards people for exceptional work.

Profit is *operating capital* that shores up money shortfalls.

Profit is *return-on-investment capital* that rewards you, the contractor-owner, for taking risks.

Without profit, an HVAC business cannot subsist, much less grow. Profit is the fuel of progress.

If a business misuses or abuses profit, however, the penalty is much like having no profit at all. Imagine the plight of an HVAC contractor who has way too much return-on-investment capital and not enough investment capital, bonus capital, and operating capital. Can you see the imbalance this creates?

The Third Kind of Money: Flow

Flow is what money *does* in an HVAC business, as opposed to what money is. Whether the business is large or small, money tends to move erratically through it, much like a pinball. One minute it's there; the next minute it's not.

Flow can be even more critical to a company's survival than profit, because a business can produce a profit and still be short of money. Has this ever happened to you? It's called profit on paper rather than in fact.

No matter how large your business, if the money isn't there when it's needed, you're threatened—regardless of how much profit you've made. You can borrow it, of course. But money acquired in dire circumstances is almost always the most expensive kind of money you can get.

Knowing where the money is and where it will be when you need it is a critically important task of both the technician-as-employee and the contractor-as-owner.

Rules of Flow

You will learn no lesson more important than the huge impact flow can have on the health and survival of your HVAC sole proprietorship, let alone your business or enterprise. The following two rules will help you understand why this subject is so critical.

1. **The First Rule of Flow states that your income statement is static, while the flow is dynamic.** Your income statement is a snapshot, while the flow is a moving picture. So, while your income statement is an excellent tool for analyzing your company *after* the fact, it's a poor tool for managing it in the heat of the moment.

Your income statement tells you (1) how much money you're spending and where, and (2) how much money you're receiving and from where.

Flow gives you the same information as the income statement, plus it tells you *when* you're spending and receiving money. In other words, flow is an income statement moving through time. And that is the key to understanding flow. It is about management in real time. How much is coming in? How much is going out? You'd like to know this daily, or even by the hour if possible. Never by the week or month.

You must be able to forecast flow. You must have a flow plan that helps you gain a clear vision of the money that's out there next month and the month after that. You must also pinpoint what your needs will be in the future.

Ultimately, however, when it comes to flow, the action is always in the moment. It's about *now*. The minute you start to meander away from the present, you'll miss the boat.

Unfortunately, few HVAC contractors pay any attention to flow until it dries up completely and slow pay becomes no pay. They are oblivious to this kind of detail until, say, customers announce that they won't pay for this or that. That gets a contractor's attention because the expenses keep on coming.

When it comes to flow, most HVAC contractors are flying by the proverbial seat of their pants. No matter how many people you hire to take care of your money, until you change the way you think about it, you will always be out of luck. No one can do this for you.

Managing flow takes attention to detail. But when flow is managed, your life takes on an incredible sheen. You're swimming with the current, not against it. You're in charge!

2. **The Second Rule of Flow states that money seldom moves as you expect it to.** But you do have the power to change that, provided you understand the two primary sources of money as it comes in and goes out of your HVAC business.

The truth is, the more control you have over the *source* of money, the more control you have over its flow. The sources of money are both inside and outside your business.

Money comes from *outside* your business in the form of receivables, reimbursements, investments, and loans.

Money comes from *inside* your business in the form of payables, taxes, capital investments, and payroll. These are the costs associated with attracting customers, delivering your services, operations, and so forth.

Few contractors see the money going *out* of their business as a source of money, but it is.

When considering how to spend money in your business, you can save—and therefore make—money in three ways:

Do it more effectively.

Do it more efficiently.

Stop doing it altogether.

By identifying the money sources inside and outside of your business, and then applying these methods, you will be immeasurably better at controlling the flow in your business.

But what are these sources? They include how you:

manage your services

buy supplies and equipment

compensate your people

plan people's use of time

determine the direct cost of your services

increase the number of clients seen

manage your work

collect reimbursements and receivables

In fact, every task performed in your business (and ones you haven't yet learned how to perform) can be done more efficiently and effectively, dramatically reducing the cost of doing business. In the process, you will create more income, produce more profit, and balance the flow.

The Fourth Kind of Money: Equity

Sadly, few HVAC contractors fully appreciate the value of equity in their HVAC business. Yet equity is the second most valuable asset any contractor will ever possess. (The first most valuable asset is, of course, your life. More on that later.)

Equity is the financial value placed on your HVAC business by a prospective buyer.

Thus, your *business* is your most important product, not your services. Because your business has the power to set you free. That's right. Once you sell your business—providing you get what you want for it—you're free!

Of course, to enhance your equity, to increase your business's value, you have to build it right. You have to build a business that works. A business that can become a true business, and a business that can become a true enterprise. A company/business/enterprise that can produce income, profit, flow, and equity better than any other contractor's business can.

To accomplish that, your business must be designed so that it can do what it does systematically and predictably, every single time.

Yes, allow me to say that again: "To accomplish that, your business must be designed so that it can do what it does systematically and predictably, every single time!"

The Story of McDonald's

Let me tell you the most unlikely story anyone has ever told you about the successful building of an HVAC company, business, and enterprise. Let me tell you the story of Ray Kroc.

You might be thinking, "What on earth does a hamburger stand have to do with my business? I'm not in the hamburger business; I'm an HVAC contractor."

Yes, you are. But by installing HVAC as you have been taught, you've abandoned any chance to expand your reach, help more customers, or improve your services the way they must be improved if the business of HVAC—and your life—is going to be transformed.

In Ray Kroc's story lies the answer.

Kroc called his first McDonald's restaurant "a little money machine." That's why thousands of franchisees bought it. And the reason it worked? Kroc demanded consistency, so that a hamburger in Philadelphia would be an advertisement for one in Peoria. In fact, no matter where you bought a McDonald's hamburger in the 1950s, the meat patty was guaranteed to weigh exactly 1.6 ounces, with a diameter of 3⅝ inches. It was in the McDonald's Operations Manual.

Did Kroc succeed? You know he did! And so can you, once you understand his methods. Consider just one part of his story.

In 1954, Kroc made his living selling the five-spindle Multi-mixer milkshake machine. He heard about a hamburger stand in San Bernardino, California, that had eight of his machines in operation, meaning it could make forty shakes simultaneously. This he had to see.

Kroc flew from Chicago to Los Angeles, then drove sixty miles to San Bernardino. As he sat in his car outside Mac and Dick McDonald's restaurant, he watched as lunch customers lined up for bags of hamburgers.

In a revealing moment, Kroc approached a strawberry blonde in a yellow convertible. As he later described it, "It was not her sex appeal but the obvious relish with which she devoured the hamburger that made my pulse begin to hammer with excitement."

Passion.

In fact, it was the French fry that truly captured his heart. Before the 1950s, it was almost impossible to buy fries of consistent quality. Kroc changed all that. "The French fry," he once wrote, "would become almost sacrosanct for me, its preparation a ritual to be followed religiously."

Passion and preparation.

The potatoes had to be just so—top-quality Idaho russets, eight ounces apiece, deep-fried to a golden brown, and salted with a shaker that, as Kroc put it, kept going "like a Salvation Army girl's tambourine."

As Kroc soon learned, potatoes too high in water content—even top-quality Idaho russets vary greatly in water content—come out soggy when fried. And so Kroc sent out teams of workers, armed with hydrometers, to make sure all his suppliers were producing potatoes in the optimal solids range of 20 to 23 percent.

Preparation and passion. Passion and preparation. Look those words up in the dictionary, and you'll see Kroc's picture. Can you envision your picture there?

Do you understand what Kroc did? Do you see why he was able to sell thousands of franchises? Kroc knew the true value of equity, and, unlike Steve from our story, Kroc went to work *on* his business rather than *in* his business. He knew the hamburger wasn't his product—McDonald's was!

So what does *your* HVAC business need to do to become a little money machine? What is the passion that will drive you to build a business that works—a turnkey system like Ray Kroc's?

Equity and the Turnkey System

What's a turnkey system? And why is it so valuable to you? To better understand it, let's look at another example of a turnkey system that worked to perfection: the recordings of Frank Sinatra.

Frank Sinatra's records were to him as McDonald's restaurants were to Ray Kroc.

They were part of a turnkey system that allowed Sinatra to sing to millions upon millions of people without having to be there himself.

Sinatra's recordings were a dependable turnkey system that worked predictably, systematically, automatically, and effortlessly to produce the same results every single time—no matter where they were played, and no matter who was listening.

Regardless of where Sinatra was, his records just kept on producing income, profit, flow, and equity, over and over . . . and still do!

Sinatra needed only to produce the prototype recording and the system did the rest.

Just like Ray Kroc's System – McDonald's thousands upon thousands of store worldwide – did and does, continuously. Long after Ray Kroc is gone!

Kroc's McDonald's is another prototypical turnkey solution, addressing everything McDonald's needs to do in a basic, systematic way so that anyone properly trained by McDonald's can successfully reproduce the same results.

And this is where you'll realize *your* equity opportunity: in the way your company does business; in the way your business systematically does what you intend it to do; and in the development of your turnkey system—a system that works even in the hands of ordinary people (and HVAC contractors less experienced than you) to produce extraordinary results.

Remember:

If you want to build vast equity *in* your business (like Ken Goodrich has and will continue to do!) then go to work *on* your business, building it into a business that works every single time just like Ken has, just like Ray Kroc has, just like every turnkey business system does, and will continue to do.

Go to work *on* your business to build a totally integrated turnkey system that delivers exactly what you promised every single time . . . no matter who's doing it!

Go to work *on* your business to package it and make it stand out from the HVAC businesses you see everywhere else.

Here is the most important idea you will ever hear about your business and what it can potentially provide for you:

The value of your equity is directly proportional to how well your business works. And how well your business works is directly proportional to the effectiveness of the systems you have put into place upon which the operation of your business depends.

Whether money takes the form of income, profit, flow, or equity, the amount of it—and how much of it stays with you—invariably boils down to this. Money, happiness, life—it all depends on how well your business works. Not on your people, not on you, but on the system.

Your business holds the secret to more money. Are you ready to learn how to find it?

Earlier in this chapter, I alerted you to the inevitable conflict between the technician-as-employee and the contractor-as-owner. It's a battle between the part of you working *in* the business and the part of you working *on* the business. Between the part of you working for income and the part of you working for equity.

Here's how to resolve this conflict:

Be honest with yourself about whether you're filling *employee* shoes or *owner* shoes.

As your company's key employee, determine the most effective way to do the job you're doing, *and then document that job.*

Once you've documented the job, create a strategy for replacing yourself with someone else (another technician, or, even better, a contractor) who will then use your documented system exactly as you do.

Have your new employees manage the newly delegated system. Improve the system by quantifying its effectiveness over time.

Repeat this process throughout your company wherever you catch yourself acting as employee rather than owner.

Learn to distinguish between ownership work and employee-ship work every step of the way.

Master these methods, understand the difference between the four kinds of money, develop an interest in how money works in your company . . . and then watch it flow in with the speed and efficiency of a perfectly pounded hammer.

Now let's take another step in our strategic thinking process. Let's look at the subject of *planning*. But first, let's see what Ken has to say about *money*. ❧

CHAPTER 4

Eyes on the Money

Ken Goodrich

Assets put money in your pocket, whether you work or not, and liabilities take money from your pocket.

—Robert Kiyosaki

E arly in my business, I often found myself short on cash. You might know what I'm talking about—not being able to pay the bills or struggling to make payroll. In the past, I worked for customers, and after making the mistake of not billing them, I never got paid for the job or wasn't able to collect on it.

In one case, I had done work for a bar in Las Vegas called Fuddy Duddy's. The bar owed me about $12,000 and never paid. It had reached the point that unless I collected it that week, I wouldn't make Friday payroll.

Fuddy Duddy's catered to casino workers who would go over after their work shifts. It was common practice for workers to cash their paychecks at casinos in order to gamble and drink. One Thursday,

I finally sat down with the owner and told her that I needed to receive at least $7,000 of the $12,000 she owed me so that I could pay my employees. She gave me her usual song and dance about why she couldn't pay the bill. I continued to be persistent about it until she said, "Okay, just wait at the bar."

I sat down, and in a mirror over the bar, I saw her ducking out the back door. I followed her out, and as soon as I opened the back door, I saw her speeding off in her Mercedes.

I felt humiliated and really pissed off. I went back to the office and had a great idea. The next day, I issued $12,000 worth of paychecks to my employees. In each envelope, I put a note to the employee, which read, "Please cash this check at Fuddy Duddy's tomorrow."

I didn't have the money in my bank account to cover it, but if they cashed the checks at her place, that would essentially force her to pay me. Monday morning, she came over to my office very unhappy.

"What you did really cramped my business," she said.

"Well, you know," I said, "now we're flush. Now we're even."

It was a harrowing experience, but I learned to always keep my eyes on the money.

In this chapter, I show you the key money-related numbers for you to keep your eyes on each day. I also tell you about the importance of these money elements: cash flow, gross profit, gross margin, and equity. If I hadn't figured these things out some thirty years ago, I never could have turned my company around, nor could I have helped people be more successful working with me or for me than they could be anywhere else. When you track your money, your company will have the cash systems to reach your financial goals, while helping others to reach theirs.

Learn to Count Your Money

Early on, I was so caught up in the doing of the work that I didn't pay a lot of attention to the money. When I finally stopped to count it, it became apparent that I'd better start paying attention. Regardless of how hard I'd worked, there wasn't enough.

I went out and bought a biography of Tom Monaghan, the founder of the Domino's Pizza franchise. In it, he said that he had failed three times and that Domino's Pizza had been his last chance. He had named his business "Domino's Pizza" because his other businesses had fallen like dominoes. The lesson he had learned was to hire three times more accountants than he ever thought he would need.

I took that suggestion to heart and immediately put together an accounting staff. We started counting *everything*. I was finally able to understand the relationship between my time and money in terms of how much I was making or not making and why. Soon after that, I put a sign up in my office that read, "Let's assume for just a moment we're in business to make money." I wanted to remind myself and all of my employees of our real purpose: to create an enterprise that generates profits.

With my team of accountants (I say "team," but it was more like three people), we set out to have good, solid daily reporting. What the numbers quickly showed us was that a business really runs on cash flow, not profits. Cash flow is an equation: the difference between the cash that comes in and the cash that goes out, every day. The result can vary from neutral to positive or negative from one day to another.

Daily Cash Flow

Cash flow was the first place I started, and I think it's the most important place for you to start. In order to succeed in this business, you need to understand your cash flow on a daily, weekly, and monthly basis. Then you have to drive your business operations to create that cash on a daily, weekly, and monthly basis.

Before the IRS came and seized my assets, they sent me a letter that said, "We want you to bring in a cash flow report so we can determine how much you can pay." I had no idea what a cash flow report was. We put the letter aside and hoped the IRS would go away.

One day, I got a call from a gentleman from California calling himself an IRS negotiator. He claimed he could help me resolve my issue with the IRS.

Okay. I have nowhere to turn, I thought. *I don't know who else to talk to.*

He showed up and convinced me that he was very experienced in negotiating tax debt with the IRS. "I'll need $5,000 to start," he said, after reviewing my bank statements.

I had $7,000 in the bank. But I was so desperate, I gave him $5,000.

Three weeks later, he showed up at my office again. He gave me a document with a cash flow report. He also asked for another $5,000. I was suspicious but desperate. I needed that report for the IRS. So I gave the guy another $5,000.

The cash flow report was a detailed week-by-week estimate of how much money had been coming into my business, how much money was going out, and what the *net cash* was. I learned that net cash was not profit; in my case, it was cash available for the single purpose of paying the IRS something every week or every month.

More importantly, he taught me what *cash flow* meant and that understanding it was a key element to running my business. He also taught me how crucial it was for cash flow that we collect payment at the time of service from customers, rather than billing them.

That negotiator didn't help me with the IRS. In fact, he provoked them to the point that they showed up with badges and guns to seize my assets. But I got far more out of that experience than he will ever know. The principles of *cash flow, immediate payment for services,* and *net cash* are what I still use to this day to run a nine-figure business. It was an expensive lesson, but I feel like I got real value from it: I learned to keep my eyes on my money.

I can't stress to you enough that a residential service and replacement business should always be based on cash on delivery. Never bill homeowners. Collect your money via checks, credit cards, or consumer financing programs. In that scenario, your residential HVAC service and replacement business has tremendous cash flow, because you're collecting your money daily but paying your bills thirty to forty-five days out. This gives you a tremendous opportunity to use the cash coming in for other things, such as growth and new strategies.

Until I understood the need for daily cash and set my business goals based on those metrics, I couldn't succeed in this business. I called it my Daily Cash Collected goal, and I rallied all my employees around it. The Daily Cash Collected was accounts receivable and cash collected from daily operations.

The daily cash flow is number one. Keep your cash flow flowing.

Monthly Cash Requirements

You also want to track your outflow of cash every day. How do you do that? Have your accounting staff put together your monthly budget—in other words, what it costs to keep the lights on. This isn't as daunting as you might think, if you break that number down by the week and, finally, by the day. You can put together your budget based on the anticipated direct costs of the work you will do that month, as well as your fixed overhead costs. For example, if you clearly understand that your rent is $3,000 a month and there's an average of thirty days in a month, you know that you need to generate $100 a day to at least cover the rent expense.

The Revenue-Generating Process

From the beginning, I always encouraged my sales management team to keep focused on this simple gross revenue-generating process:

1. Lead generation: Buy and/or create sales leads for new air conditioners or service calls.
2. Lead conversion: Book the lead and sell the lead.
3. Client fulfillment: Get the work done and collect the money.

This may sound very simple, but there is no reason to over-complicate your business. Together, these steps form a system that generates consistent sales.

Key Drivers to Revenue

When the dispatcher sends a technician on an HVAC service call, the technician should come back with one of four results. Each result generates one of four types of revenue, as follows:

1. Diagnostic fee—The technician diagnoses the problem but provides no other service. The diagnostic fee covers the cost of the technician, wages, and vehicle, but the visit does not generate actual profit.
2. Flat-rate repair fee—The technician finds a problem, performs a repair, and generates a flat-rate repair fee.
3. Replacement sale—The technician determines that the product is no longer functional or is beyond economical repair and recommends a system replacement. This generates a replacement fee. This revenue source has the highest yield.
4. Service agreement—The customer purchases a service or maintenance contract, which creates recurring revenue. It also creates an ongoing relationship with the customer. I talk about this in more detail in Chapter 14.

The sales drivers fill each of those four sources with the gross profit dollars that they've created in each of these opportunities. That, in turn, covers your overhead, and the spillover creates net profit.

Balancing the Cash Flow: Gross Profit and Gross Margin

After you clearly understand the cash coming in and going out on a daily basis—cash flow—you can focus on your *gross profit*. These are the profit dollars that are a result of the work the company performs each day. This is calculated by taking your total sales less all direct costs associated with providing the services, such as the materials, labor, insurance, credit card fees, and financing fees.

The percentage that every business owner should keep an eye on is the *gross margin*. Gross margin is determined by dividing the gross profit by the total sales. This will give you a percentage. To run a successful operation, your minimum standard should be a 42 percent gross margin. For a healthy and growing HVAC service and replacement business, the benchmark is closer to 50-plus percent gross margin.

The key mission of the HVAC contractor is to develop a business that creates healthy profits. Exceptional players in our industry can achieve as much as a 60 percent gross margin by maintaining the sales price and controlling their costs.

The 50/30/20 Benchmark

The prototypical model for the HVAC service and replacement business is a 50 percent gross margin, with a 30 percent overhead. From these percentages, you can determine your net profit: a 50 percent gross margin less a 30 percent overhead results in a 20 percent net profit. I've seen companies that produce net profits in excess of 30 percent, but a good benchmark to start working toward is 50 percent less 30 percent, which equals 20 percent net profit.

Quality, Not Quantity

If you follow the rule of 50 percent gross margin, maintain your overhead at 30 percent, and work toward a 20 percent net profit, your pricing methodology will fall in line. That being said, you cannot worry about what the other guy is charging. In the HVAC service and replacement business, most contractors are limited by capital, people, large-scale experience, and lack of other resources.

I've seen a lot of guys get in trouble by comparing their HVAC contracting businesses to Walmart. They say, "Well, look at Walmart. They're the biggest and cheapest, and they do it on volume." Here's what you need to know: Walmart has countless big warehouses full of inventory its vendors put there and constantly replenish. In our business, people buy an air conditioner maybe once or twice in a lifetime. People buy Walmart goods twice a week. Unlike Walmart, our business is not a volume game. It's a competence game.

Hands down, the leader in every major HVAC market is the premium-priced and premium service provider. This is not so much the contractor aiming to be the most expensive, it's really a result of these contractors understanding what their costs are. The majority of people just starting out in our business do not understand. Most of us who evolved from service technicians to business owners learned how to price our products and services by asking the guy at the parts house counter what he thinks. It's absolutely the wrong thing to do.

There's an art and a science to pricing. It's based on knowing your true costs and building a healthy profit. You can learn more from organizations such as the Service Nation Alliance, Service Roundtable, and ACCA. These organizations can help you understand how to properly price your products and services.

The HVAC contractor who is most competent and takes care of his or her customers best owns a market share, not the contractor who is the cheapest. If a contractor benchmarks at 50 percent, 30 percent, and 20 percent, he'll get a great return on his business and start to build the next part of the money, which is equity.

The Value of Equity

Once you've mastered cash flow and you've got your arms around profit, your next focus has to be equity. Predictable gross margins, controllable overhead costs, and predictable profitability create the most valuable businesses. The definition of equity that Michael and I use is the enterprise value of the business.

Your main responsibility as the CEO of your company is to maximize shareholder value. In other words, your main focus is to build an enterprise that produces profits without your daily involvement and is valued in multiples of the profits that it generates. A good rule of thumb for the valuation of an HVAC service and replacement business is between four and eight times annual profits. Enterprise value, aka equity, is one of the vital factors the CEO should be keenly focused on.

To Sell or Not to Sell

Many people tell me they're never going to sell their business. But, ultimately, everyone will sell their business at some point. They're going to sell it to the bankruptcy court, a creditor, a family member, a strategic buyer, or even a competitor. That's because we don't live forever, and we certainly don't want to work forever.

In my opinion, a strategic buyer (private equity or a regional or national company) will always yield the highest sale price, because such a buyer has the capital to buy companies for top dollar and the resources to grow them. Knowing that your business will get sold one way or another should motivate you to think about equity, allowing your business to be sold for the highest and best price.

Every business can be likened to a three-legged stool. Money is just one leg of the stool. The other two legs we will talk about later in this book. But at some point, if you're too hyper-focused on money, it will slow you down. When you have diverse groups of people working for you, you soon realize not everyone is motivated by money. Of course, people need to make a living and support their families. But that's not the only reason they come to work at your company every day; nor should it be yours. We have to be conscious of the money—count it, be respectful—but, again, it's not the only leg on that stool.

In the next chapter, Michael tells you about the Planning Triangle and how to plan who you are, what you do, and how you will do it. ❧

CHAPTER 5

On the Subject of Planning

Michael E. Gerber

People in an organization operating from a creative mode . . . approach planning first by determining what they truly want to create, thus in essence becoming true to themselves.

— Robert Fritz, *The Path of Least Resistance*

Another obvious oversight revealed in Steve and Peggy's story was the absence of true planning.

Every contractor starting his or her own business must have a Plan. You should never begin to see customers without a Plan in place. But, like Steve, most contractors do exactly that.

An HVAC contractor lacking what I call "a Dream, a Vision, a Purpose and a Mission" which is the overriding Strategic Plan of your company, is simply someone who goes to work every day. Someone who is just "doing it, doing it, doing it. Busy, busy, busy." Maybe making money, maybe not. Maybe getting something out of his or her life, maybe not. But what we can absolutely know for sure is that any such contractor is taking chances without really taking control.

The Plan tells anyone who needs to know *how we do things here*.

The Plan defines the objective and the process by which you will attain it.

The Plan encourages you to organize tasks into functions, and then helps people grasp the logic of each of those functions.

This in turn permits you to bring new employees up to speed quickly.

There are numerous books and seminars on the subject of heating and cooling, but they focus on the technical aspects of making you a better HVAC contractor. I want to teach you something that you've never been taught before: how to be a Manager. How to be Leader.

As you'll discover, this has nothing to do with thinking like an entrepreneur.

The Planning Triangle

As we discussed in the Preface, every HVAC sole proprietorship is a business, every HVAC business is a company, and every HVAC company is an enterprise.

The trouble with most companies owned by an HVAC technician is that they are dependent on the technician.

That's because they're a sole proprietorship—the smallest, most limited form a company can take. Businesses that are formed around the expert technician, whether HVAC installer or repairman, always struggle to grow, and, most often, as we've said over and over again, relentlessly, fail to do it, and then, as night turns into day, fail to wake up one morning, sooner than later.

You may choose in the beginning to form a sole proprietorship, but you should understand its limitations. The company called a *sole proprietorship* depends on the owner—that is, the technician. The company called a *business* depends on other people plus a system by which that business does what it does. Once your sole proprietorship becomes a business, you can replicate it, turning it into an *enterprise*.

Consider the example of Sea HVAC. The customers don't come in asking for Douglas Sea, although he is one of the top HVAC contractors around. After all, he can only handle so many calls a day and be in only one location at a time.

Yet he wants to offer his high-quality services to more people in the community. If he has reliable systems in place—systems that any qualified associate contractor can learn to use—he has created a business and it can be replicated. Douglas can then go on to offer his services—which demand his guidance, not his presence—in a multitude of different settings. He can open dozens of HVAC businesses, none of which need Douglas Sea himself, except in the role of entrepreneur.

Is your HVAC company going to be a sole proprietorship, a business, or an enterprise? Planning is crucial to answering this all-important question. Whatever you choose to do must be communicated by your Plan, which is really three interrelated plans in one. We call it the Planning Triangle, and it consists of:

The Business Plan

The Service Plan

The Completion Plan

The three plans form a triangle, with The Business Plan at the base, The Service Plan in the center, and The Completion Plan at the apex.

The
Completion
Plan

The Service Plan

The Business Plan

The Business Plan determines *who* you are (the business). The Service Plan determines *what* you do (the specific focus of your HVAC business) and The Completion Plan determines *how* you do it (the fulfillment process).

By looking at the Planning Triangle, we see that the three critical plans are interconnected. The connection between them is established by asking the following questions:

1. *Who are we?* Purely a strategic question

2. *What do we do?* Both a strategic and a tactical question

3. *How do we do it?* Both a strategic and a tactical question

Strategic questions shape the vision and destiny of your business, of which your service is only one essential component. Tactical questions turn that vision into reality. Thus, strategic questions provide the foundation for tactical questions, just as the base provides the foundation for the middle and apex of your Planning Triangle.

First ask: What do we do, and how do we do it *strategically?*

And then: What do we do, and how do we do it *practically?*

Let's look at how the three plans will help you develop our service.

The Business Plan

Your business plan will determine what you choose to do in your HVAC company and the way you choose to do it. Without a business plan, your company can do little more than survive. And even that will take more than a little luck.

Without a business plan, you're treading water in a deep pool with no shore in sight. You're working against the natural flow.

I'm not talking about the traditional business plan that is taught in business schools. No, this business plan reads like a story—the most important story you will ever tell.

Your business plan must clearly describe

the business you are creating

the purpose it will serve

the vision it will pursue

the process through which you will turn that vision into a reality

the way money will be used to realize your vision

Build your business plan with *business* language, not *technical* language (the language of the HVAC contractor). Make sure the plan focuses on matters of interest to your lenders and shareholders rather than just your technicians. It should rely on demographics and psychographics to tell you who buys and why; it should also include projections for return on investment and return on equity. Use it to detail both the market and the strategy through which you intend to become a leader in that market, not as an HVAC sole proprietorship but as a business enterprise.

The Business Plan, though absolutely essential, is only one of three critical plans every contractor needs to create and implement. Now let's take a look at the three legs of The Service Plan.

The Service Plan

The Service Plan includes everything an HVAC contractor needs to know, have, and do in order to deliver his or her promise to a customer on time, every time.

Every task should prompt you to ask three questions:

1. What do I need to know?
2. What do I need to have?
3. What do I need to do?

What Do I Need to Know?

What information do I need to satisfy my promise on time, every time, exactly as promised? In order to recognize what you need to know, you must understand the expectations and limitations of others, including your customers, administrators, technicians, and other employees. Are you clear on those expectations? Don't make the mistake of assuming you know. Instead, create a need-to-know checklist to make sure you ask all the necessary questions.

A need-to-know checklist might look like this:

What are the expectations of my customers?

What are the expectations of my technicians?

What are the expectations of my staff?

What are the expectations of my vendors?

What Do I Need to *Have?*

This question raises the issue of resources—namely, money, people, and time. If you don't have enough money to finance operations, how can you fulfill those expectations without creating cash-flow problems? If you don't have enough trained people, what happens then? And if you don't have enough time to manage your business, what happens when you can't be in two places at once?

Don't assume that you can get what you need when you need it. Most often, you can't. And even if you can get what you need at the last minute, you'll pay dearly for it.

What Do I Need to *Do?*

The focus here is on actions to be started and finished. What do I need to do to fulfill the expectations of this customer on time, every time, exactly as promised? For example, what exactly are the steps to ensure the customer has a comfortable and consistent temperature in every room of the home?

Your customers fall into distinct categories, and those categories make up your business. The best HVAC businesses will invariably focus on fewer and fewer categories as they discover the importance of doing one thing better than anyone else.

Answering the question *What do I need to do?* demands a series of action plans, including:

the objective to be achieved;

the standards by which you will know that the objective has been achieved;

the benchmarks you need to read in order for the objective to be achieved;

the function/person accountable for the completion of the benchmarks;

the budget for the completion of each benchmark; and

the time by which each benchmark must be completed.

Your action plans become the foundation for the Completion Plan. And the reason you need Completion Plans is to ensure that everything you do is not only realistic but can also be managed.

The Completion Plan

If The Business Plan gives you results and provides you with standards, The Completion Plan tells you everything you need to know about every benchmark in The Service Plan—that is, how you're going to fulfill customer expectations on time, every time, as promised. In other words, how you're going to ensure the customer's airflow is consistent in every room, set a thermostat for optimal energy saving, or educate a customer about the importance of regularly servicing their ducts and filters.

The Completion Plan is essentially the operations manual, providing information about the details of doing tactical work. It is a guide to tell the people responsible for doing that work exactly how to do it.

Every Completion Plan becomes a part of the knowledge base of your business. No Completion Plan goes to waste. Every Completion Plan becomes a kind of textbook that explains to new employees or new associates joining your team how your business operates in a way that distinguishes it from all other HVAC businesses.

To return to an earlier example, The Completion Plan for making a Big Mac is explicitly described in the *McDonald's Operation Manual*, as is every Completion Plan needed to run a McDonald's business.

The Completion Plan for an HVAC technician might include the step-by-step details of how to measure a customer's airflow using a digital manometer at every vent—in contrast to how everyone else has learned to do it. Of course, all those who work in HVAC have used a digital manometer. They've learned to do it the same way everyone else has learned to do it. But if you are going to stand out as unique in the minds of your customers, employees, and others, you must invent your own way of doing even ordinary things. Most of that value-added perception will come from your communication skills, your listening skills, and your innovative skills in transforming an ordinary visit into a customer experience.

Perhaps you'll decide that a mandatory part of airflow analysis is to sketch a quick map of airflow cfm (cubic feet per minute) at every vent and show it to the customer, explaining what the different numbers mean so that she has a better understanding of her home. If no other contractor your customer has seen has ever taken the time to explain how to balance airflow, you'll immediately set yourself apart. You must constantly raise the questions from: *How do we do it here?* To the more elevating question: *How should we do it here?*

The quality of your answers will determine how effectively you distinguish your business from every other HVAC contractor's business.

Benchmarks

You can measure the movement of your business—from what it is today to what it will be in the future—using business benchmarks.

These are the goals you want your business to achieve during its lifetime.

Your benchmarks should include the following:

Financial benchmarks

Emotional benchmarks (the impact your business will have on everyone who comes into contact with it)

Performance benchmarks

Customer benchmarks (Who are they? Why do they come to you? What will your business give them that no one else does?)

Employee benchmarks (How do you grow people? How do you find people who want to grow? How do you create a school in your business that will teach your people skills they can't learn anywhere else?)

Your business benchmarks will reflect (1) the position your company will hold in the minds and hearts of your clients, employees, and investors; and (2) how you intend to make that position a reality through the systems you develop.

Your benchmarks will describe how your management team takes shape and what systems you need to develop so your managers, just like McDonald's managers, will be able to produce the results for which they will be held accountable.

Benefits of the Planning Triangle

By implementing the Planning Triangle, you will discover such things as:

What your business will look, act, and feel like when it's fully evolved

When that's going to happen

How much money you will make

These, then, are the primary purposes of the three critical Plans: (1) to clarify precisely what needs to be done to get what the contractor wants from his or her business and life, and (2) to define the specific steps by which it will happen.

First *this* must happen, then *that* must happen. One, two, three. By monitoring your progress, step-by-step, you can determine whether you're on the right track or not.

That's what planning is all about. It's about creating a standard—a yardstick—a Benchmark, against which you will be able to measure your performance.

Failing to create such a standard is like throwing a straw into a hurricane. Who knows where that straw will land?

Have you taken the leap? Have you accepted that the words *business* and the word *sole proprietorship* are not synonymous? That a sole proprietorship relies on the contractor as the first and foremost technician at hand to do the work that needs to be done, and a business relies on other people who are trained to do that work exactly as it MUST be done, along with the most essential part of this equation, the absolutely essential set of SYSTEMS which tell everyone, this is how we do it here!?

Because most contractors are control freaks—indeed, most technicians are—99 percent of today's HVAC companies are sole proprietorships, not businesses.

The result, as a friend of mine says, is that "contractors are spending all day stamping out fires when all around them the forest is ablaze. They're out of touch. And to get IN touch that contractor better take control of the business they've created before someone else does."

And I say that because of the irrefutable fact that the vast majority of contractors of whatever kind are never taught to think like business people, the skilled tradesperson is forever at war with the business-person. This is especially evident in large HVAC companies, where bureaucrats (who are commonly thought to be business people_ often try to control contractors ("technicians suffering from an entrepreneurial seizure").

Unfortunately, rather than each getting better at what each are supposed to do, they usually end up treating each other as combatants.

In fact, the single greatest reason contractors become entrepreneurs is to divorce themselves from such bureaucrats and begin to reinvent the HVAC enterprise.

That's you.

And now that the divorce is over and a new love affair has begun, you're an HVAC contractor with a Plan!

Who wouldn't want to do business with such a person?

Now let's take the next step in our strategic odyssey. Let's take a closer look at the subject of *management*. But before we do, let's read what Ken has to say on the subject of *planning*. ❧

6

Plan
Your Dreams

Ken Goodrich

Unless commitment is made, there are only promises and hopes; but no plans.

—Peter Drucker

From age sixteen, I dreamt of running many successful businesses. But it wasn't until I read *The E-Myth Revisited*™ that I understood that unless I wrote down my dreams, they would never become a plan or a road map to reality. I am happy to say that I have accomplished almost everything I have ever written down as a goal.

I was forced to plan out of sheer frustration and anger. After the IRS seized my assets and slapped on fines, my business had no available credit. Most of the vendors I dealt with would not even take my checks. Every dollar I collected went straight back into the company. The $3,000 cash I had saved in a suit pocket was my working capital. I no longer had my BMW, having traded it in at the Ford dealership for an F-150 pickup with only an AM radio, rubber floor mats, and a ladder rack.

Each morning at 6:30 a.m., I showed up at Allied Refrigeration to buy parts for jobs sold the previous day. One bright and early morning, I arrived to make my usual purchase. The manager, Derek, approached me with a sour look on his face.

"You can't come by here at 6:30 in the morning anymore," he said. "We need to get the orders ready for my *paying* customers."

"What do you mean, *paying* customers? I'm paying you in cash every day."

"I'm talking about our *important* customers."

"I'm not an important customer to you?" I snapped.

Infuriated and embarrassed, I retreated to my truck with my hands clenched. I picked up *The E-Myth Revisited*™ along with a yellow notepad. Right there in the parking lot, I wrote these words on the pad: "My strategic objective is to sell this business within five years for $1 million." At the time, I thought that was all the money in the world.

To fulfill my goal, as Michael said, I knew I needed to create a clear plan. I continued writing: "I need to have systems that my employees can follow to deliver good service to my customers. I need to get good people and train them in those systems. I need to be the premium provider in the marketplace, so I can afford good marketing to attract new customers every day. I need to do the very best work so customers will come back and refer me. I have to build a profitable business that operates without me so someone will want to buy it." From there, I created a detailed road map to reach my goal.

Within five years of writing my plan, I accumulated three companies and sold them for more than six times my original goal. Plans are powerful. They turn your dreams into a road map for success.

In this chapter, I talk about how to start your plan, measure your success, and set goals. Like Michael says, if the only one who knows your plan is you, you're just dreaming. A plan is a documented road map to share with your team. Providing a clear, transparent path gives you a way to achieve your dreams while helping others in your business to succeed. To get what you want, you have to give people what they want.

Where to Start Your Plan

You have to start with the end in mind, by visualizing your business in the future. What is your business going to look like when it is running perfectly to your standards? When I started my first plan, that meant my business was creating the amount of profits I desired and running smoothly without my day-to-day involvement to build a valuable and saleable asset. It also meant that it was thriving ethically and morally. My goal was to build a company that would provide opportunities to every person who joined my team and worked hard. Once you have your end goal in mind, you backtrack and create a road map on how to get there.

Your road map can be as simple as what I started out with: one page of a yellow legal pad. That one page transformed my dreams into a solid plan. It can be as simple as "I'm going to build an HVAC service and replacement business that produces $2 million in annual net profit and has an enterprise value of at least $10 million." Then you start listing the resources you need in terms of people, profit, and your pricing model, marketing strategy, and client fulfillment process.

On that yellow notepad, I wrote what the business was going to look like to the customers, what it was going to mean to the employees, and what kind of employees I needed in my company to reach my end goal, which back then was to sell the business.

I put the plan up on my office wall and looked at it every single day until I internalized it. I constantly checked in with myself and asked, "Am I heading in that direction or not?" That reminder, plus the fact that everyone in my office saw it and held me accountable to it, made it a concrete, solid plan.

The business plan I wrote was executed every single day after that. I ran the business during the day and documented our business systems at night. I rallied my team around improving our company and helping to build systems.

It seemed like pie in the sky at the time, but I told them, "The systems we are building today are going to positively influence the entire industry in the future. What we build here today is going to be used across the country and will improve the lives of thousands of HVAC professionals. We are now working for a higher purpose than our paychecks."

None of this was easy. Dedicated people stayed with me even during the hardest times, because they believed I could provide an opportunity for them and their family.

By the time I finished, we had several volumes of operating manuals. New employees could be trained on how to consistently provide the premium quality my business would be known for. The documents clearly explained how we priced our products, filled out invoices, answered the telephones, and installed our air conditioning systems, along with every other aspect of the business.

The goal was to duplicate myself without having to manage every detail on my own.

Plans and Timing

Once you have your initial overall plan, you can decide how much time it's going to take you to achieve it. If your overall plan has a ten-year timeline, you'll break it down into ten one-year plans. Your annual business plan should project your revenue, gross profit, gross margin, and net profit. It can also list new products and services you plan to offer, along with your lead generation, lead conversion, and client fulfillment strategies for the year.

After that first year, your road maps from this point forward should reflect on the past (where your company has been), the present (where your company is now), and the future (where your company needs to go).

Enterprise Value Strategy

Just as I looked at the plan tacked on my wall, you and your ʌʌ⸝ must keep a keen focus on whether or not you're building something of value—an *enterprise value strategy*. Enterprise value is created by having predictable and growing profits in your business. A simple method to keep your team focused on the future within your ten-year plan is to describe the profitability of each of your years, multiplied by four to eight, and what the value of your company will look like in each of those ten years.

Vital Metrics

We have many vital metrics by which we measure our business, and yours might be different, but here are some of ours:

Demand calls by month

Maintenance calls by month

Diagnostic fee only

Maintenance fee only

Demand-to-repair ratio

Maintenance-to-repair ratio

Demand turnover ratio

Maintenance turnover ratio

Replacement sale closing ratio

Replacement sale average ticket

Cost per lead

Call center booking rates

Gross margin

Net margin

Revenue

Net income

The list is not exclusive. To have a good plan, you need key performance indicators of how your business is doing, to plan what your business will do in the future.

Track Your Progress

It is imperative that you and your team routinely track your progress in following the plan. For instance, in our operations, our team conducts two vital factors calls per week, whereby we review our essential metrics, discuss any challenges, and share best practices to stay on track. Monthly, we have a key management team vital factors meeting as taught to us by Management Action Programs (MAP) of Newport Beach, CA (more about this in Chapter 8). In this meeting, we review our successes over the past month, team members who should be recognized for over-and-above performance, current challenges in the business, and our list of vital factors or metrics.

Once we identify key challenges from the prior month that prevent the execution of our plan, we clearly define the challenges, agree on who will fix them, and provide a deadline to do so. Each month, we score ourselves as a team by how many fixes we decided we were going to fix and how many we actually completed. We divide the completed fixes by the actual number of challenges to get a team completion score. On our team, our goal is to fix our monthly challenges at the rate of at least 85 percent. This keeps the business on course with our plans, in order to improve and grow.

The Goal System

Even though you have a well-documented plan, it does not mean that your people are going to automatically follow it. You have to motivate them to follow the plan. After I made my plan, I went back to *The E-Myth* book and read that you should motivate your people with a game. I created a goal system as a method for positive rewards.

For every employee in the company, I created a scorecard similar to a baseball card with the metrics by which we measured the success of our plan and statistics that indicated how well each person was sticking to the plan. In the course of this, I realized that we had some home run hitters who sold the big jobs, but we also had some good runs batted in (RBI) players who took good care of customers, making them happy and loyal to the company. Then we had other players who were great at selling service contracts. This was a system where every player was equal and could win by doing the job he or she excelled at—in other words, his or her contribution to the company's plan.

I would treat the teams that achieved their goals—for example, hitting the cash-flow goal—to a Goal Dinner. We would get dressed up in suits and go out to a fine dining restaurant with a private room where they could experience high-end customer service. It was also a time for them to feel rewarded and appreciated for hitting their goals.

As soon as I initiated the goal system, two things happened. One, I had the most cohesive, driven, energetic team in the industry, and we began to take the market. We didn't miss our cash collected goal for sixty-two months straight. Two, I had more time to do the work of innovating, documenting, and implementing business systems to catapult our business.

Your plan should not be about growing incrementally each year. I encourage you to define your true potential and your team's true potential each and every year and grow accordingly.

I came up with a motto for the company and everyone who worked with us: *If we can't count it, we don't do it.* If you think you need to plan for something, but you can't quantify the return on it, then don't do it.

Take off the chains. See how far you can fly. There's no reason why your plan can't be about taking huge leaps forward, instead of taking incremental baby steps. Write it down, rally your team around it, and own it.

In the next chapter, Michael talks about how managers don't manage people as much as they manage processes. ❧

CHAPTER 7

On the Subject of Management

Michael E. Gerber

The most important figures that one needs for management are unknown or unknowable, but successful management must nevertheless take account of them.

—W. Edwards Deming

Every contractor, including Steve Walsh from our story, eventually faces the issue of management. Most face it badly.

Why do so many contractors suffer from a kind of paralysis when it comes to management? Why are so few able to get their business to work the way they want it to and run it on time? Why are their managers (if they have any) seemingly so inept?

Why do so many contractors suffer from a kind of paralysis when it comes to dealing with management? Why are so few able to get their HVAC business to work the way they want it to and to run it on time? Why are their managers (if they have any) seemingly so inept?

There are two main problems. First, the contractor usually abdicates accountability for management by hiring an office manager. Thus, the contractor is working hand in glove with someone who is supposed to do the managing. But the contractor is unmanageable himself!

The contractor doesn't think like a manager because he doesn't think he is a manager. He's an HVAC contractor! He rules the roost. And so he gets the office manager to take care of stuff like scheduling appointments, keeping his calendar, collecting receivables, hiring/firing, and much more.

Second, no matter who does the managing, they usually have a completely dysfunctional idea of what it means to manage. They're trying to manage people, contrary to what is truly needed.

We often hear that a good manager must be a "people person." Someone who loves to nourish, figure out, support, care for, teach, baby, monitor, mentor, direct, track, motivate, and, if all else fails, threaten or beat up his or her people.

Don't believe it. Management has far less to do with people than you've been led to believe.

In fact, despite the claims of every management book written by management gurus (who have seldom managed anything), no one—with the exception of a few bloodthirsty tyrants—has ever learned how to manage people.

And the reason is simple: People are almost impossible to manage.

Yes, it's true. People are unmanageable. They're inconsistent, unpredictable, unchangeable, unrepentant, irrepressible, and generally impossible.

Doesn't knowing that make you feel better?

Now you understand why you've had all those problems! Do you feel the relief, the heavy stone lifted from your chest?

The time has come to fully understand what management is really all about. Rather than managing *people*, management is really all about managing a *process*, a step-by-step way of doing things, which, combined with other processes, becomes a System.

For example:

The Process for on-time scheduling

The Process for answering the telephone

The Process for greeting a customer

The Process for organizing customer files

Thus, a Process is the step-by-step way of doing something over time. Considered as a whole, these processes aggregate into a System:

The On-time Scheduling System

The Telephone Answering System

The Customer Greeting System

The File Organization System

Instead of managing people, then, the truly effective manager has been taught how to manage a series of Systems for managing a Process through which People get things done.

More precisely, managers and their people, *together,* manage the Processes—which are constructed of the Systems which comprise the Operating Reality of your business.

Management is less about who gets things done in your business than about how things get done.

In fact, great managers are not fascinated with people, as our contemporary mantra suggests they must be, but instead with how things get done through people using extraordinarily effective Systems to do it.

To do that, great managers constantly ask themselves and their people key questions, such as:

What is the result we intend to produce?

Are we producing that result every single time?

If we're not producing that result every single time, why not?

If we are producing that result every single time, how could we produce even better results?

Do we lack a system? If so, what would that system look like if we were to create it?

If we have a system, why aren't we using it?

And so forth.

In short, a great manager can leave the office every day fully assured that it will run at least as well as it does when he or she is physically in the room.

Great managers are those who use a great management system. A system that shouts, "This is *how* we manage here." Not, "This is *who* manages here."

In a truly effective company, how you manage is always more important than who manages.

Provided a system is in place, how you manage is transferable, whereas who manages isn't.

How you manage can be taught, whereas *who* manages can't be.

When a company is dependent on *who* manages—Katie, Kim, or Kevin—that business is in serious jeopardy. Because when Katie, Kim, or Kevin leaves, the business has to start over again.

What an enormous waste of time and resources!

Even worse, when a company is dependent on *who* manages, you can bet all the managers in that business are doing their own thing.

What could be more unproductive than ten managers who each manage in a unique way? How in the world could you possible manage those managers?

The answer is: You couldn't.

Because it takes you right back to trying to manage *people* again.

And, as I hope you now know, that's impossible.

In this chapter, I often refer to managers in the plural.

I know that most HVAC contractors only have one manager—the Office Manager.

And so you may be thinking that a Management System isn't so important in a small HVAC business. After all, the Office Manager does whatever an Office Manager does (and thank God because *you* don't want to do it). Why go through the trouble of creating a Management System?

But if your sole proprietorship is ever going to turn into the Business it could become, and if that Business is ever going to turn into the Enterprise of your dreams, then the questions you ask about how the Office Manager manages your affairs are critical ones.

Because until you come to grips with your dual role as owner and key employee, and the relationship your manager has to those two roles, your sole proprietorship/business/enterprise will never realize its potential. Thus the need for a Management System.

Management System

What, then, is a Management System?

The E-Myth says a Management System is the method by which every manager innovates, quantifies, orchestrates, and then monitors the systems through which your service produces the results you expect.

According to *The E-Myth*, a manager's job is simple:

A manager's job is to invent the Systems through which the owner's vision is consistently and faithfully manifested at the operating level of the business.

Which brings us right back to the purpose of your business and the need for an entrepreneurial vision.

Are you beginning to see what I'm trying to share with you? That your business is *one single thing?* And that all the subjects we're discussing here—money, planning, management, and so on—are all about doing one thing well?

That one thing is the one thing your service is intended to do: distinguish your HVAC business from all others.

It is the manager's role to make certain it all fits.

And it's your role as entrepreneur to make sure your manager knows what the business is supposed to look, act, and feel like when it's finally done.

As clearly as you know how, you must convey to your manager what you know to be true—your vision, your picture of the business when it's finally done. What I called earlier, "Your Dream, Vision, Purpose and Mission."

In this way, your vision is translated into your manager's marching orders every day he or she reports to work.

Unless that vision is embraced by your manager, you and your people will suffer from the tyranny of routine. And your business will suffer from it, too.

Now let's move on to people. Because, as we know, it's people who are causing all our problems! But first let's read what Ken has to say about management. ✤

The Skilled Trade of Management

Ken Goodrich

What gets measured, gets managed.

—Peter Drucker

Have you ever wished a Prince Charming would ride up on his white horse and save your business? I did.

When I was twenty-eight, I decided to hire my first manager. I was about two years into rebuilding my business. The business wasn't quite successful yet, but things had calmed down with the IRS, and I was sticking to my plan and cash flow goals. I needed to free up my time to work on my growing business, so I put an ad in the paper.

I received a reply from a retired vice president of a big mechanical contracting company. I was enthusiastic about his credentials. After we met, he was excited about the residential service replacement business cash flow model I showed him. I was thinking, *I'm going to get this guy into my business and he's going to make me rich!*

The pay he required was beyond anything I had ever paid before—in fact, it was more money than even I made. However, I thought if he moved the business forward, he would be worth the money.

When he reported to work, he asked to be left alone so he could survey the business to get a good understanding of how things ran. I assumed after he completed this exercise, he would make his recommendations, so I immediately agreed. Unfortunately, I didn't have the foresight to add a timeline.

He bought himself a big whiteboard and some sticky notes. Every day, he would stand in front of his board, writing on the notes and putting them on the whiteboard. He would talk to our people at long lunches, take more notes, and add more sticky notes to the big whiteboard. He carried on like this for four months. Finally, getting impatient with his huge drain on my finances, I asked, "Okay, what's the plan?"

"Come in tomorrow," he said, "and I'll show you my plan with the sticky notes."

The next day, I sat down at our meeting and was excited. This former vice president of a big mechanical contractor and his sticky notes were going to save me. *He's going to give me the keys to the kingdom, I thought. We're going to take off together, and I'm going to have the greatest air conditioning company in Las Vegas!*

After his presentation, two things became clear. One, he had nothing to offer me with the sticky notes and whiteboard. Two, after going back to *The E-Myth Revisited*™, I realized I had hired and paid a guy thousands of dollars to write a system on my business, which he knew nothing about. It turned out, he wasn't retired at all. He had been terminated from the big mechanical contractor for poor performance. That day, history repeated itself as he was terminated from my company.

Here's what I learned: Prince Charming is never going to ride up on a white horse to your front door and say, "I'm here to manage your business and make you millions." Your managers utilize the systems you give them to manage your business.

In this chapter, I talk about management as a system that allows you to grow rapidly. I also share the methods we use in our company, including the five principles of management that we use.

I can't be your Prince Charming, but you can use the principles here to create your own system. Our management system is about creating managers within our system who will build our business. When you have a management system, you have processes that cultivate your people to grow and become better at what they do.

Managers Are Made

Management systems are the systems to manage people. I used to think good managers are either the nice guy who cajoles employees, or the hard-ass who pushes. What I learned is managers need to know more about managing systems and less about managing people. If managers can oversee the system, the only people-managing they ever have to do is to make sure their people operate within the system.

I also thought good managers had to be the smartest people in the room. When I chose a manager, I would ask myself: Who is my best technician? Or, who is my best salesperson? I ended up making the mistake of promoting my best technician or salesperson to manager. The result was I took my best people out of the field, and, generally speaking, they would turn out to be poor managers, because they were in unfamiliar roles. It's like I cut off my nose to spite my face.

What most new business owners don't understand is management is as much of a skill as being an HVAC technician. It took us years to become great HVAC technicians, so how could we expect employees to become talented managers overnight? Management is another skilled trade that requires a methodology. Just as an air conditioner is a set of processes, management is an organized, step-by-step process that functions to produce a result, in this case, a smooth-running business.

Our managers utilize the Five Fundamentals of MAP from MAP Consulting. These fundamentals ensure that documented business are being followed and measured within our company. Founded in 1960, MAP Consulting is a pioneer of business management consulting and executive/leadership training. Prior to learning the MAP principles, my management teams utilized many systems, where we managed people and not systems. With MAP, we learned to develop our management team and systems. Their methodology provided our greatest tools to reach our goals and stay on plan.

The Five Fundamentals of MAP

According to the Five Fundamentals of MAP, a company must provide its managers with (1) a business plan, (2) a management system, (3) job descriptions, (4) timely and meaningful performance appraisals, and (5) ninety-day training plans. These tools provide the essentials for our managers to drive and motivate their teams, which in turn grows our business. A more detailed breakdown of each MAP Fundamental is as follows.

Fundamental #1: A Business Plan

It is important to understand where your business has been, where it is, and where it is going. MAP refers to this as your company's road map. It is also important to understand how your company will get to the next level.

If you, your managers, and your employees know this information, everyone visualizes the company's goals and is knowledgeable about the path to reaching them. Your business plan should be shared with your management team and managers routinely, so they can all understand where the business is heading and how it is getting there.

Fundamental #2: Management System

According to MAP, the goal of a management system is to ensure your company is living and executing your plan. The easiest way to determine if you are executing your plan is to measure your key performance indicators (KPIs) or Vital Factors®. KPIs are the vital metrics by which we drive and measure the business's performance.

During weekly and monthly check-ins, management holds their teams accountable to ensure everyone is staying on plan. Managers also measure whether goals were met by each team member, in alignment with the company's road map, or the business plan.

In this area, managers also engage their team members in team consulting, where they document and solve any challenges the company is facing. Managers sit down with their teams and do a mini-discovery, or an analysis of what's working, what's not working, and why. Then they brainstorm to solve those problems.

Fundamental #3: Job Descriptions

MAP suggests that each position in the company have a job description. In my company, we use *position agreements*, which are like a next-level job description. A position agreement describes the job title, the work to be done, the result of the work to be done, the strategic work for that job, the tactical work of the job, and how that job gets measured.

For example, one clear expectation we have for customer service reps is that they will make great first impressions with the customers when they schedule their service calls. Whether you use a job description or a position agreement, it is crucial that the expectations of each position are clearly explained in the document, as these expectations will be used in the next factor.

Fundamental #4: Performance Appraisals

Employees must be given feedback and direction. The proper way to do this is through timely, meaningful performance appraisals. By using the expectations clearly listed in the job description or position agreement, a manager can properly perform a review of the employee and provide meaningful feedback and direction in addition to the KPIs that will also be measured for that employee.

These appraisals ideally should take place every quarter, or annually at the very least. Performance appraisals are vital parts of the business that frequently are overlooked because managers feel too busy and don't make them a priority. Performance appraisals make a solid foundation for your company. Managers can use the time to talk about the strengths and weaknesses of each employee and give the employee tools to grow with professional development plans.

Fundamental #5: Ninety-Day Onboarding/Training Plan

When we first started the business, the extent of our onboarding was to ask new technicians a few questions about how to fix an air conditioner and then throw them the keys to a van. We set everyone—ourselves, the company, and the employees—up for failure by not including a ninety-day onboarding plan with goals. Now, all new hires undergo this training.

Every onboarding plan also needs to include a mentor, appointed by managers. Mentors informally show new hires the ropes—in other words, not just where to get your coffee or where the restrooms are, but what the company culture is like. They also fill new hires in on little nuances that you won't find in the manual, such as whom to talk to for this or that.

Prince Charming won't come to save you. But by creating a management system, you can become your own Prince Charming. Management isn't a natural ability; it can be learned and perfected just like any other skill.

In the next chapter, Michael shifts our focus to people by proving the People Law: "Without people, you don't own a business, you own a job." ❧

On the Subject of People

Michael E. Gerber

*When you innovate, you've got to be prepared for people telling you
that you are nuts.*

—Larry Ellison, founder of Oracle Corporation

Every HVAC contractor I've ever met has complained
about people.

About employees: "They come in late, they go home
early, they have the consistency of an antique boiler!"

About property managers: "They're living in a nonparallel universe!"

About customers: "They want to keep their heater made in 1979!"

People, people, people. Every contractor's nemesis. And at the
heart of it all are the people who work for you.

"By the time I tell them how to do it, I could have done it
twenty times myself!" "How come nobody listens to what I say?"
"Why is it nobody ever does what I ask them to do?"

Does this sound like you?

So what's the problem with people? To answer that, think back to the last time you walked into an HVAC contractor's office. What did you see in the employee's faces?

Most people working in HVAC are harried. You can see it in their expressions. They're negative. They're tired. They're humorless. And with good reason! After all, they're surrounded by people who have clogged ducts, outdated pipes, or—worst-case scenario—may even be breathing carbon monoxide! Customers are looking for the most value for the least amount of money. And many are either angry or frightened. They don't want any problems with their homes.

Is it any wonder employees at most HVAC businesses are disgruntled? They're surrounded by unhappy people all day. They're answering the same questions 24/7. And most of the time, the owner has no time for them. He or she is too busy leading a dysfunctional life.

Working with people brings great joy—and monumental frustration. And so it is with owners and their employees. But why? And what can we do about it?

Let's look at the typical HVAC contractor—who this person is and isn't.

Most contractors are unprepared to use other people to get results. Not because they can't find people, but because they are fixated on getting the results themselves. In other words, most contractors are not the businesspeople they need to be, but *technicians suffering from an entrepreneurial seizure*.

Am I talking about you? What were you doing before you became an entrepreneur?

Were you an HVAC technician working for a large commercial contractor? A subcontractor working for a midsized residential builder? Or just doing odd jobs here and there?

Didn't you imagine owning your own business as the way out?

Didn't you think that because you knew how to do the technical work—because you knew so much about airflow and ventilation—that you were automatically prepared to create a business that does that type of work?

Didn't you figure that by creating your own business you could dump the boss once and for all? How else to get rid of that impossible person, the one driving you crazy, the one who never let you do your own thing, the one who was the main reason you decided to take the leap into a business of your own in the first place?

Didn't you start your own business so that you could become your own boss?

And didn't you imagine that once you became your own boss, you would be free to do whatever you wanted to do—and to take home *all* the money?

Honestly, isn't that what you imagined? So you went into business for yourself and immediately dived into work.

Doing it, doing it, doing it.

Busy, busy, busy.

Until one day you realized (or maybe not) that you were doing all of the work. You were doing everything you knew how to do, plus a lot more you knew nothing about. Building sweat equity, you thought.

In reality, an HVAC technician suffering from an entrepreneurial seizure.

You were just hoping to make a buck in your own business. And sometimes you did earn a wage. But other times you didn't. You were the one signing the checks, all right, but they went to other people.

Does this sound familiar? Is it driving you crazy?

Well, relax, because we're going to show you the right way to do it this time.

Read carefully. Be mindful of the moment. You are about to learn the secret you've been waiting for all your working life.

The People Law

It's critical to know this about the working life of contractors who own their own HVAC business: *Without people, you don't own a company, you own a job.* And it can be the worst job in the world because you're working for a lunatic! (Nothing personal—but we've got to face facts.)

Let me state what every contractor knows: Without employees, you're going to have to do it all yourself. Without human help, you're doomed to try to do too much. This isn't a breakthrough idea, but it's amazing how many contractors ignore the truth. They end up knocking themselves out, ten to twelve hours a day. They try to do more, but less actually gets done.

The load can double you over and leave you panting. In addition to the work you're used to doing, you may also have to do the books. And the organizing. And the filing. You'll have to do the planning and the scheduling. When you own your own business, the daily minutiae are never-ceasing—as I'm sure you've found out. Like painting the Golden Gate Bridge, it's endless. Which puts it beyond the realm of human possibility. Until you discover how to get it done by somebody else, it will continue on and on until you're a burned-out husk.

But with others helping you, things will start to drastically improve. If, that is, you truly understand how to engage people in the work you need them to do. When you learn how to do that, when you learn how to replace yourself with other people—employees trained in your system—then your business can really begin to grow. Only then will you begin to experience true freedom yourself.

What typically happens is that contractors, knowing they need help answering the phone, filing, and so on, go out and find people who can do these things. Once they delegate these duties, however, they rarely spend any time with the hoi polloi. Deep down, they feel it's not important *how* these things get done; it's only important *that* they get done.

They fail to grasp the requirement for a system that makes employees their greatest asset rather than their greatest liability. A system so reliable that if Chris dropped dead tomorrow, Leslie could do exactly what Chris did. That's where the People Law comes in.

The People Law says that each time you add a new person to your business using an intelligent (turnkey) system that works, you expand your reach. And you can expand your reach almost infinitely! People allow you to be everywhere you want to be simultaneously, without actually having to be there in the flesh.

Employees are to an HVAC contractor what a record was to Frank Sinatra. A Sinatra record could be (and still is) played in a million places at the same time, regardless of where Frank was. And every record sale produced royalties for Sinatra (or his estate).

With the help of other people, Sinatra created a quality recording that faithfully replicated his unique talents, then made sure it was marketed and distributed, and the revenue managed.

Your employees can do the same thing for you. All you need to do is to create a "recording"—a system—of your unique talents, your special way of installing and servicing HVAC, and then replicate it, market it, distribute it, and manage the revenue.

Isn't that what successful businesspeople do? Make a "recording" of their most effective ways of doing business? In this way, they provide a turnkey solution to their customers' problems. A system solution that really works.

Doesn't your business offer the same potential for you that records did for Sinatra (and now for his heirs)? The ability to produce income without having to go to work every day?

Isn't that what your employees could be for you? The means by which your system for installing and servicing HVAC could be faithfully replicated?

But first you've got to have a system. You have to create a unique way of doing business that you can teach to your employees, that you can manage faithfully, and that you can replicate consistently, just like McDonald's.

Because without such a system, without such a "recording," without a unique way of doing business that really works, all you're left with is people doing their own thing. And that is almost always a recipe for chaos. Rather than guaranteeing consistency, it encourages mistake after mistake after mistake.

And isn't that how the problem started in the first place? Employees doing whatever they perceived they needed to do, regardless of what you wanted? Employees left to their own devices, with no regard for the costs of their behavior? The costs to you?

In other words, employees without a system.

Can you imagine what would have happened to Frank Sinatra if he had followed that example? If every one of his recordings had been done differently? Imagine a million different versions of "My Way." It's unthinkable.

Would you buy a record like that? What if Frank was having a bad day? What if he had a sore throat?

Please hear this: The People Law is unforgiving. Without a systematic way of doing business, people are more often a liability than an asset. Unless you prepare, you'll find out too late which ones are which.

The People Law says that without a specific system for doing business; without a specific system for recruiting, hiring, and training your employees to use that system; and without a specific system for managing and improving your systems, your business will always be a crapshoot.

Do you want to roll the dice with your business at stake? Unfortunately, that is what most contractors are doing.

The People Law also says that you can't effectively delegate your responsibilities unless you have something specific to delegate. And that something specific is a way of doing business that works!

Sinatra is gone, but his voice lives on. And someone is still counting his royalties. That's because Sinatra had a system that worked.

Do you?

Now we will move on to the subject of *subcontractors*. But before we do, let's see what Ken has to say about *employees*. ❧

CHAPTER 10

Harness the Talent

Ken Goodrich

When people are financially invested, they want a return. When they are emotionally invested, they want to contribute.

—Simon Sinek, author of *Start with Why*

One day, as I was trying to figure out how to find the time to create systems to run the business while running the business itself, a guy walked into my office. His name was Curtis Coker. He told me that he had just retired as a sergeant from the Air Force, knew HVAC, and needed a job. I like to hire military people because they tend to be well trained, disciplined, and reliable. So, I hired Curt.

Curt really took to our goal system, which I wrote about in Chapter 6. He made sure the system was followed to a T. He was the most well-rounded player on the team.

At the same time, I was growing more frustrated, because I couldn't move the business forward. I was still too involved in the day-to-day operations. I thought, *All I'm doing when I come into work every day is prolonging my agony.*

I called Curt into my office and said, "Curt, I want you to be the general manager of this company. I can't pay you much, but as we grow, there'll be more opportunity and I can pay you what you're worth. But right now, I need the help so that I can go in and start working on more systems to help the company move forward. Otherwise, we're just going to be stuck right here."

He agreed.

I centered his position around the key things that I needed to have happened: manage the fieldwork and collect the daily cash so the business would stay afloat, while I created the systems. I paid Curt a salary to be the general manager. Plus, if we hit our monthly cash flow goal, he would get a monthly bonus of a percentage of the positive cash flow.

When you have a people system, your people can work on your business and perform better. Meanwhile, you are freed up to work on your business. In this chapter, I talk about hiring people, creating systems for your people to follow, rewarding hard work, and the importance of positive company culture. Having a business system that includes recognition and reward helps people perform better and thrive. When your people thrive, so does your business.

The Biggest Challenge

Keeping employees happy is critical to your top and bottom line. One of the most expensive challenges in our industry is employee turnover. It can cost tens of thousands of dollars per person to recruit, rehire, and retrain a new employee after losing one, not to mention the lost revenue while the position is not filled. I have found employee turnover costs $18,000 per person, at a minimum, and could cost tens of thousands more if you lose a key salesperson. Creating a people retention system is paramount to the success of your enterprise.

You may have seen advertising for open jobs in the trades, as we are desperate for skilled labor. The trades are a great way to make a living, as we know, but sitting in an air-conditioned office and going to lunch with your peers may sound a little more glamorous than climbing on a roof or in an attic to work on air conditioners.

That makes your people system even more critical. It is essential to do everything you can to retain those who are qualified to work for your company.

The First Big Hire

People are your greatest resource for your success, and you must honor them and treat them accordingly. When you take that big step to hire your first employee, remember the care and attention you gave that experience, and from here on out, hold all your new hires in the exact same regard.

Your workplace is a community that your employees will engage in for most of their day. Coming to work, for them, is like entering into another community in their lives. Think about it, their neighborhoods, kids' schools, and sports teams are also communities they belong to. These communities need to keep them engaged and committed, and they must be fun. They are places where they can grow and flourish and feel safe, respected, and rewarded. Your business is no different; in fact, it's more so. It's a big responsibility.

The foundation of your people system is your company's values. For instance, at Goettl, our company's values are excellence, action, growth, and innovation. When you get started in your business, sit down and list your community, or company, values. Use that as your guiding light in all things, including when you hire and when you fire. Even when you let someone go, you can refer to your values as the reason. When someone does not meet or match the company values, that's when you need to part ways.

To be successful, I had to figure out a way to run my business through people rather than doing it all myself. As Michael says, if I did it myself, I would have only so much time and, therefore, I could make only a limited amount of money. I started to hire people, but until I put them into systems, I didn't have time to manage them!

One of my first hires should have been a human resources manager. The reason why I never invested in a good HR team was because I always looked at HR as the hall monitors for a business—the people who nitpicked about compliance—rather than a value-add to the business. What I've learned is that HR is one of the most important roles in a company, in terms of finding the right people for the job and for the company culture. They put people in the right seats on the bus and develop and grow them to be increasingly beneficial to the company and to their own careers.

Unconscious Bias in Hiring

Not everyone wants what I want; nor will everyone put in the same amount of work as I, the owner of the company, will. That was a hard lesson for me to learn. At first, I hired people with unconscious bias. They were just like me. While I have my strengths, I certainly also have my weaknesses. I'm not best suited for every role in the company. At one point, I found myself surrounded by people who could do everything I could do, but not the stuff I needed them to do.

You're going to have people on your team who want to be the CEO one day or the top salesperson. You're also going to have people who want to clock in, clock out, and go home. Either way, you want to have employees who want to be part of a good, solid company. Each one of those types is equally necessary and vital to the success of the business, and each type requires a different system to operate by.

The System Is the Solution

Next, I hired people for their technical skills, and I sent them out to do the right thing by the customers and the company. But I had limited results on that. Then I started to hire people for their character and communication skills. I sent them out, hoping they would do the right thing by the customers and the company, and I had limited success with that, too.

After learning and implementing the E-Myth principles, I started to create systems. I found that *both* the technically skilled employees and the employees with strong communication skills had equally great success after following my system. The solution was not in the people, but in the system.

Cycles of Development

One of my favorite sayings from Michael is "Your business is a school for employees." That's why my company's HR team includes a director of training and development.

My company is a training ground for employees to learn the basic functions of their jobs and pursue personal and professional growth. It's also a school that teaches employees about our brand and our whys—why we do what we do and are a part of something that's bigger than your typical day-to-day grind.

You and your HR team must be able to discuss your employees in a frank and open, yet strictly confidential, manner in terms of their performance, aptitude, and development. This is not gossiping or talking behind your employees' back in a negative way. This is about productive, positive discussions about the employees with the goal of helping them professionally develop and be more of an asset to the company.

I regularly have discussions with my team about each one of our key management people, to examine the directions in which they are heading. For example, when one employee aspired to be COO, we resolved to develop him toward that role. We told him that we were going to invest in him and put him through a cycle of development within our business that would lead him toward that goal. As a result, this manager became more engaged and even more committed to the business. He saw that we were investing in him, and he started stepping up even more. Employees must be engaged within a cycle of development based on their goals.

Our human resources team has come up with a list of questions to ask key managers so we can decide what their futures look like in the company. Here are some:

How engaged do you feel you are in the day-to-day business at Goettl?

What can we do to help you prepare for life after Goettl?

What skills, experiences, and behaviors would benefit you to learn in the next two years that can benefit you and the company?

If you could design any role at Goettl to maximize your engagement and contribution in the next two years, what would it be?

The answers give us the direction we need to give our employees a growth path for success.

Positive Company Culture

With my HR team, I started to work on the people system for my company. The people system is about creating a positive culture inside the business. My company culture is about *achievement, recognition*, and *reward*. We are all a part of something bigger than ourselves; we are a part of a great, winning team. That gets people excited to come to work every day.

Compensation ranks number three or four on most employees' priority lists. Number one, for most employees, is recognition. We've also created a recognition system inside the business to honor our employees' exceptional performance. We've created quarterly *employee appreciation outings*, in the form of *goal dinners*, movies, picnics, or departments getting together for karaoke. We encourage and fund those kinds of activities to connect people with our culture and brand.

Earlier, I told you about the goal dinners we have for our employees when we obtain a certain goal. For our employees, the real reward isn't the fancy dinner, but the recognition and appreciation. Throughout the dinner, I or a key manager will make a point to mention each employee in a positive manner.

For instance, using their monthly scorecards, I may say, "Joe is a great part of this team because when Mrs. Jones was very upset about our service, he turned that around. Now she's a champion for our company, and she referred us to two other neighbors. We're so lucky to have Joe on the team."

Camaraderie forms during the dinners. The dinners also motivate everyone to go after more goals. The next Monday, employees often come back to work and say, "All right, we got to hit this one. We've got to have another goal dinner."

As Michael says, a company without the right people is nothing. Hiring key players gave our company skills and talents that I alone never could have. When Curt took over as GM more than thirty years ago, he ran the field better than I could. He was more empathetic with our people. He got them to work harder than I could. He created more discipline in the company than I ever could because I was always anxious about not having time to create the systems.

With Curt as GM, our business flourished. I pumped out the systems as fast as I could, and he implemented and followed them religiously. In the process, Curt's salary increased by five times. Curt Coker still works with me today.

In the next chapter, Michael talks about a subject that gives most HVAC contractors the biggest headaches: *subcontractors*, and how to manage them. ✤

On the Subject of Subcontractors

Michael E. Gerber

Meaningful relationships and meaningful work are mutually reinforcing, especially when supported by radical truth and radical transparency.

—Ray Dalio

I f you're a sole practitioner—that is, you're selling only yourself—then your HVAC company, called a sole proprietorship, will never make the leap to an HVAC business called a company. The progression from sole proprietorship to business to enterprise demands that you hire other technicians to do what you do (or don't do). In some cases, these people might be subcontractors.

Subcontractors are different from regular employees in all states. They operate under different laws and regulations as owners of their solely-owned company. No matter what name they operate under, they too, must operate within your System. To the degree they don't, you're in trouble.

Contractors know that subs can be a huge problem. Until you face this special business problem, your sole proprietorship will never become a business, and your business will certainly never become an enterprise.

Long ago, God said, "Let there be contractors. And so they never forget who they are in my creation, let them be damned forever to hire people exactly like themselves." Enter the subcontractors.

Solving the Subcontractor Problem

Let's say you're about to partner with a subcontractor. Someone who has specific skills: electrical, ducting, ventilation, whatever. It all starts with choosing the right personnel. After all, these are subcontractors to whom you are delegating your responsibility and for whose behavior you are completely liable. Remember Frank Sinatra? Do you really want to leave that choice to chance? Are you that much of a gambler? I hope not.

If you've never worked with your subcontractor, how do you really know he or she is skilled? For that matter, what does "skilled" mean?

For you to make an intelligent decision about this subcontractor, you must have a working definition of the word *skilled*. Your challenge is to know exactly what your expectations are, then to make sure your subcontractors operate with precisely the same expectations. Failure here almost always assures a breakdown in your relationship.

I want you to write the following on a piece of paper: "By *skilled*, I mean . . ." Once you create your personal definition, it will become a standard for you and your business, for your customers, and for your subcontractors.

A standard, according to *Webster's Eleventh*, is something "set up and established by authority as a rule for the measure of quantity, weight, extent, value, or quality."

Thus, your goal is to establish a measure of quality control, a standard of skill, which you will apply to all your subcontractors. More important, you are also setting a standard for the performance of your company.

By creating standards for your selection of subcontractors—standards of skill, performance, integrity, financial stability, and experience—you have begun the powerful process of building a company that can operate exactly as you expect it to.

By carefully thinking about exactly what to expect, you have already begun to improve your business.

In this enlightened state, you will see the selection of your subs as an opportunity to define what you (1) intend to provide for your customers, (2) expect from your employees, and (3) demand for your life.

Powerful stuff, isn't it? Are you up to it? Are you ready to feel your rising power?

Don't rest on your laurels just yet. Defining those standards is only the first step you need to take. The second step is to create a *Subcontractor Development System.*

A Subcontractor Development System is an action plan designed to tell you what you are looking for in a subcontractor. It includes the exact benchmarks, accountabilities, timing of fulfillment, and budget you will assign to the process of looking for subcontractors, identifying them, recruiting them, interviewing them, training them, managing their work, auditing their performance, compensating them, reviewing them regularly, and terminating or rewarding them for their performance.

All of these things must be documented—actually *written down*—if they're going to make any difference to you, your subcontractors, your managers, or your bank account!

And then you've got to persist with that System, come hell or high water. Just as Ray Kroc did at Mc Donald's. Just as Walt Disney did at Disneyland. Just as Sam Walton did at Walmart.

This leads us to our next topic of discussion: the subject of *estimating.* But first, let's read what Ken has to say on the subject of *subcontractors.* ❧

.

CHAPTER 12

Valued and Valuable Subcontractors

Ken Goodrich

Corporations must embrace the benefits of cooperating with one another.
—Simon Mainwaring

W hen you sell a business, the new owners often require that you do not start a new company in the same industry near the company you sold. This non-compete agreement typically lasts two to five years. In 2001, after having fulfilled the non-compete agreement from the sale of my first three businesses, I began to build a new HVAC company. I put together a new management team and bought a few small, underperforming businesses in Las Vegas and Phoenix. I branded them all as Yes! Air Conditioning.

Our slogan at Yes! was "The Yes! man can," and our brand promise was "We are capable, professional and will do whatever it takes to make you (the customer) happy." With our brand message and new industry look, the company took off. The small companies we purchased and combined grew ten times their former revenue in the first year.

However, rapid growth comes with obstacles. Finding qualified HVAC installers is always a challenge, but try doing that when your business is doubling every sixty days!

One day, while talking with our sheet metal subcontractor, Bill Moore, I asked, "Bill, what do you think about your guys installing HVAC systems for us?"

"Well, Ken," Bill said in his slow, slightly Southern drawl. "I think we can do that for ya. You just tell me how you want the jobs done, I guess."

We were off to the races with an innovation to our installation labor challenge. We delivered our documented HVAC system installation manual/system to Bill, negotiated a price, and began doing what everyone else in the industry said couldn't be done: we subcontracted out our installations.

From the very beginning, the subcontract system was flawless. We took the time to document our installation system and standards, and Bill and his team followed them perfectly. Even though the local industry scoffed at our partnership, we eventually grew to be the market leader in Las Vegas and successfully subcontracted all our installations to $12 million in revenue, when Bill decided he had made enough money and decided to retire. I bought Bill's company, Sunset Sheet Metal and Air Conditioning, and it became the installation department of a now–$30 million company.

There's no better feeling than doing something that people said couldn't be done.

In this chapter, I tackle subcontractors, the HVAC contractor's biggest headache, and explain how they can be your greatest ally, rather than the biggest thorn in your side. By holding our subcontractors to a higher standard of professionalism, we help them become more successful, as they help us become more successful.

To Subcontract or Not?

It's never a good idea to pay your employees—the people who report to you day-to-day—as subcontractors. (The IRS has a twelve-point test for employees versus subcontractors if you're unclear about the difference. You can find it on the IRS website.) If you hire your technicians or installers as subcontractors, for instance, you have less control over the quality and timeliness of their service. For example, do they share your core values?

Don't hire your technicians as subcontractors, because to some extent you can't control them. You should build an enduring, successful company with your own team members. Your subcontractors are people ancillary to the day-to-day business operations.

In HVAC, we often bring in subcontractors, such as crane operators, sheet metal fabricators, electricians, and plumbers, to name a few. Because we cannot perform work outside the scope of our licenses, we call in the experts to deliver the highest premium service to our customers. These subcontractors will not be direct employees of your company and should be vetted to ensure their reputation is as respected as yours.

Hire for Value, Not Price

We already established in Chapter 4 that, to be the market leader, you will more than likely be the premium provider in the space. I sometimes find my fellow contractors aligning themselves with subcontractors who do not have the same values, who are not the premium providers in their space. In other words, my fellow contractors hire subcontractors at the lowest price.

It's a double standard. We want our customers to base their relationships with us on high value, not price. Yet, we often do not do the same with our subcontractors. There's no way to make this work; it'll always come back to bite you. If I sell a service based on value, but I bring in a subcontractor based on price, the subcontractor is not going to deliver on our promise to the customer. And that makes me look bad. That's not to say that you must pay a premium for your subcontractors. But, if you have a value-based business, make sure that you bring in subcontractors who operate within your core values.

Vetting Your Subs

One of the mistakes that a lot of contractors make is to assume that if someone owns a business or a subcontracting business, that person must be competent enough. As you follow the *E-Myth* journey, you will realize that it's incumbent upon you to make sure that the people you do business with do business the way that you do business. Do they understand what you now know? Here are some of the key elements that you need to address with your subcontractors: Do they share your core values? Do they operate similarly to you in terms of having solid business processes to ensure the quality and timeliness of their work?

Subcontractor Agreements

After you choose your subcontractors, you need to have annual subcontractor agreements in place. These agreements should specify the terms and conditions of how your relationship is going to work for up to a year. A lawyer should draw up an annual subcontractor agreement. The agreement should outline things such as insurance requirements, workers' compensation requirements, timeliness, notice, quality control, what happens in the event of a callback, what happens in the event of an accident, payment terms, pricing, and overall ethics. Be sure to include yourself and the customer as "additionally insured" on the subcontractor's liability insurance policies as well. For every subcontractor you have, you need to have a contract in place.

Keep Communication Open

Make time to routinely meet with your key people and your subcontractors, just to iron out any challenges and make your work together more efficient. For example, you might plan weekly meetings to discuss open projects.

Streamline the Estimating Process

Once you have your subcontractor in place with an annual agreement and the pricing is already set, put that information into your price and your estimating model. When you estimate a job, you don't have to go back to subs for their estimates because all of the prices will have been previously agreed upon between you and your selected subcontractors. The prices have already been negotiated. You don't want to have to enter into new negotiations every time you use a sub with whom you have an established relationship. The subcontractors will happily train your employees to estimate for them at the flat-rate price spelled out in their annual agreement, to save them time as well.

Subcontractors don't have to give HVAC contractors headaches. By vetting them to ensure you share the same values and creating a system they can follow to align with your goals, you can bring them into your company as trusted partners. If you help subcontractors to become more successful by supporting their business and bringing more work their way, your organization will, in turn, become more successful.

In the next chapter, Michael tackles a subject that every contractor struggles to get right: *pricing* and *estimating*. By putting a precision systems solution in place, you can erase words such as "ballpark" from your vocabulary. Read on to find out how. ✣

CHAPTER 13

On the Subject
of Estimating

Michael E. Gerber

You can't manage what you can't measure.

—Peter Drucker

One of the greatest weaknesses of HVAC contractors is accurately estimating how long jobs will take and then scheduling their customers accordingly. *Webster's Collegiate Dictionary* defines estimate as "a rough or approximate calculation." Anyone who has worked on a jobsite knows that those estimates can be rough indeed.

Do you want to hire someone who gives you a rough approximation?

What if your doctor gave you a rough approximation of your medical condition?

The fact is, we can predict many things we don't typically predict. For example, there are ways to calculate for common problems. Look at the steps of the process. Most of the things you do are standard, so develop a step-by-step system and stick to it.

In my book *The E-Myth Manager*, I raised eyebrows by suggesting that doctors eliminate the waiting room. Why? You don't need it if you're always on time. The same goes for an HVAC company. If you're always on time, then your customers don't have to wait.

What if a contractor made this promise: on time, every time, as promised, or we pay for it.

"Impossible!" contractors cry. "Each job is different. We simply can't know how long each will take."

Do you follow this? Since contractors believe they're incapable of knowing how to organize their time, they build a business based on lack of knowing and lack of control. They build a business based on estimates.

I once had a contractor ask me, "What happens when we discover that a customer's ductwork needs to be replaced? How can we deal with something so unexpected? How can we give proper service and stay on schedule?"

My first thought was that it's not being dealt with now. Few contractors are able to give generously of their time. Ask anyone who's been to a contractor's office lately. It's chaos.

The solution is interest, attention, analysis. Try detailing what you do at the beginning of a job, what you do in the middle, and what you do at the end. How long does each take? In the absence of such detailed, quantified standards, everything ends up being an estimate, and a poor estimate at that.

However, an HVAC business organized around a system has time for proper attention. It's built right into the system.

Too many contractors have grown accustomed to thinking in terms of estimates without thinking about what the term really means. Is it any wonder many HVAC businesses are in trouble?

Enlightened contractors, in contrast, banish the word estimate from their vocabulary. When it comes to estimating, just say no!

"But you can never be exact," contractors have told me for years. "Close, maybe. But never exact."

I have a simple answer to that: *You have to be.* You simply can't afford to be inexact. You can't accept inexactness in yourself or in your HVAC business.

You can't go to work every day believing that your business, the work you do, and the commitments you make are all too complex and unpredictable to be exact. With a mindset like that, you're doomed to run a sloppy ship. A ship that will eventually sink and suck you down with it!

This is so easy to avoid. Sloppiness—in both thought and action—is the root cause of your frustrations.

The solution to those frustrations is clarity. Clarity gives you the ability to set a clear direction, which fuels the momentum you need to grow your business.

Clarity, direction, momentum—they all come from insisting on exactness.

But how do you create exactness in a hopelessly inexact world? The answer is this. You discover the exactness in your business by refusing to do any work that can't be controlled exactly.

The only other option is to analyze the market, determine where the opportunities are, and then organize your business to be the exact provider of the services you've chosen to offer.

Two choices, and only two choices: (1) Evaluate your business and then limit yourself to the tasks you know you can do exactly, or (2) start all over by analyzing the market, identifying the key opportunities in that market, and building a business that operates exactly.

What you cannot do, what you must refuse to do, from this day forward, is to allow yourself to operate with an inexact mindset. It will lead you to ruin.

Which leads us inexorably back to the word I have been using throughout this book: Systems.

Who makes estimates? Only contractors who are unclear about exactly how to do the task in question. Only contractors whose experience has taught them that if something can go wrong, it will—and to them!

I'm not suggesting that a systems solution will guarantee that you always perform exactly as promised. But I am saying that a systems solution will faithfully alert you when you're going off track, and will do it before you have to pay the price for it.

In short, with a systems solution in place, your need to estimate will be a thing of the past, both because you have organized your business to anticipate mistakes, and because you have put into place the system to do something about those mistakes before they blow up.

There's this, too: To make a promise you intend to keep places a burden on you and your site supervisors to dig deeply into how you intend to keep it. Such a burden will transform your intentions and increase your attention to detail.

With the promise will come dedication. With dedication will come integrity. With integrity will come consistency. With consistency will come results you can count on. And results you can count on mean that you get exactly what you hoped for at the outset of your business: the true pride of ownership that every contractor should experience.

This brings us to the subject of *customers*. Who are they? Why do they come to you? How can you identify yours? And who *should* your customers be? But first let's listen to what Ken has to say about *estimating*. ❧

CHAPTER
14

Priced Right

Ken Goodrich

Value can mean a price. Value can mean exclusivity. Value can mean "I can't get it anywhere else, and this is something I really want."
—Mindy Grossman

L eonardo, a young HVAC entrepreneur with $20 million in annual sales, had the coolest social media posts. Every day, he posted inspiring messages about success, such as a picture of Mount Everest with a quote about climbing to your highest potential. He even posted videos of himself at his air conditioning business in front of his troops, motivating them with his platitudes. He had a ton of TV interviews, media publicity, and social media followers.

When I saw his stuff on social media, I thought, *Wow, this guy seems like he's really got a good thing going.*

I eventually met this entrepreneur at a marketing event. We became friends, and he asked if I would give him some feedback on his business. He confided in me: he was five years into his business with $20 million in annual sales, but he had never made a profit.

His problem, as I saw it after asking more questions, was his pricing model. The wrong pricing model will negatively affect every part of your business—you, the employees, the customers, the subcontractors, the vendors—and it will ultimately put you out of business.

In this chapter, I show you how establishing solid pricing methodologies and prices takes a lot of frustration and fear out of our employees when they communicate with customers. The right pricing system makes their jobs easier and more productive, establishes trust and consistency with your customers, and ultimately makes your company more profitable.

Gross Profit per Crew Day Pricing

When I sat down with Leonardo, I discovered that he—like most contractors, including myself before E-Myth—was running on cash flow, not profit. As you know, cash flow is the money coming in and out of your business. Profit is the money your company keeps after everyone else, including Uncle Sam, is paid.

I told Leonardo that, as long as you're bringing in enough cash flow to pay your weekly obligations, you can survive forever. But this to me is like business purgatory because you're trapped in doing it, without a way to break free from the business. In my opinion, it's the closest to hell that you can possibly be while remaining alive.

Leonardo had hired an empty suit CFO—someone with an MBA, but no experience—to help run his business. "That's lunacy," I told Leonardo. "You're letting people who don't know your business decide your pricing. You need to base your price on your costs, and your plan."

Leonardo's idea of pricing—like most contractors' idea of pricing—was asking the counter guy at the supply house what he thinks they should charge for something. Or comparing the prices to what everyone else is charging. Or, they use the price that the last guy they worked for used.

As you know from Chapter 4, these are poor practices. In our type of business, there are very specific methodologies for pricing your products and services. A great book that helped me was *The Power of Positive Pricing* by Matt Michel. The book is a classic instruction manual on pricing in the service industry.

It is a common mistake to believe the pricing for replacing an HVAC system is based on some imaginary market that exists in your area. You are not selling a commodity such as pork bellies; you're selling an engineered home-comfort system. More specifically, you're selling a system that is custom-designed for one customer, one home at a time, based on individual needs and wants.

In addition to your customer's needs and wants, your pricing is based on other very specific factors. The first is the quality of the equipment and installation methods used in delivering that solution to your customer. Nothing is more important than the caliber of the installation when selecting a comfort solution for your home. The caliber of installation determines everything from the level of comfort experienced in the home to the energy usage, the annual cost of ownership, and overall level of client satisfaction.

The second factor is the level of service your team can provide after the installation. The ability to service the needs of your customers over the life of the system is what sets the professional service companies apart from amateurs. The average life of an HVAC business is generally far less than the average life of a system. The value experienced by eliminating long-term risk and hassle for a client always exceeds investment amount when explained to a rational consumer.

The third factor is the financial component. The two biggest aspects of this are your overhead costs and profit motive. Simply put, if you plan to make a profit of $25,000 in one month, and your overhead is $50,000 that month, you'll have to create $75,000 in gross profit that month. Said another way, if you know you can sell fifteen systems in a particular month, your average gross profit per crew day needs to be at least $5,000.

To the extent that you can provide a superior product and service, your gross profit per crew day can increase—creating more profit and fewer jobs being necessary to meet your profit motive.

Another consideration for your pricing—and an indicator of proper pricing and installation execution—is the gross margin percentage. Some contractors price on gross margin alone, but I do not recommend it.

Let's get into the math here. This may seem tedious, but it will make the difference between surviving and thriving in your business. For example, let's work with a total job cost of $5,000. The equipment is $3,500 and installation is $1,500. Adding a markup of 1.67 to achieve a 42 percent gross margin creates a sale price to the customer of $8,350.

However, using the gross profit per crew day example we just discussed, the job would contribute only $3,350 in gross profit dollars ($8,350 less $5,000). That amount is $1,650 less than your $5,000 gross profit per crew day target that is required to earn your budgeted net profit dollars of $25,000 that month.

There must be a balance between gross profit dollars and gross margin percentage when it comes to pricing. That balance will depend on many factors, such as lead generation, lead conversion, and installation efficiency. Your gross profit contribution per crew day should be used to ensure your overhead is covered and net profit goal is achieved.

Your gross margin percentage is an indicator of proper pricing. It ensures that you are not over- or undercharging. The larger and more efficient your company becomes in all areas, the more flexibility you have to become the dominant player.

Mastering the balance between gross margin percentage and gross profit dollars per crew day can be used to your benefit and provide you a competitive advantage. Until that day, I suggest you price based on the number of jobs you can sell each month, then factor in the overhead required to provide the level of service you desire and your profit goal.

Flat-Rate Service Pricing

Back in the old days of mom-and-pop contracting, small business owners came up with a market hourly rate that had no relationship to the cost of doing business. Most everyone used a similar rate based on nothing more than tribal knowledge. It was just a random number, which really kept our industry down for a long time. Service contractors could never charge enough to grow.

Flat-rate pricing was a departure from hourly pricing in the contracting industry, and that has really allowed our businesses to earn profit to grow. You need to cover the hourly cost of doing business, coupled with a productivity factor and a profit margin target, to create your hourly rate.

If you combine your calculated hourly rate with materials, you come up with a flat rate for every job that covers your direct costs, overhead, and profit. There are several methods to calculate your service department hourly rate to create your flat-rate pricing, and most flat-rate pricing systems will do that for you. In addition, serviceroundtable.com has a simple yet effective hourly rate calculator for its members that works well.

Pricing for Smaller Businesses

No matter where you are in the HVAC business, you want to price based on your current conditions: overhead, staffing levels, and the speed and agility by which you can operate. When you're new to the HVAC business, don't make the mistake of thinking that if you have low overhead, you don't need to charge as much as everyone else.

Early on in your business, it's beneficial to be more premium-priced *because* you're not as efficient. You don't have a lot of people to help you bring in dollars, but you need those profits, aka capital, to grow your business. It's seemingly counterintuitive. The small guys do not have the luxury of charging less if they truly want to grow and succeed. They must position themselves with a level of service and pricing on par with the big guys, if they want to grow.

Pricing in the Four Types of Revenue Sources

In Chapter 4, I wrote about the four types of revenue sources in an HVAC contracting business: diagnostic fees, flat-rate repair fees, replacement sales, and service agreements. You want to price at all of these.

Diagnostic Fees

A diagnostic fee, that is, the fee you charge to come out to a property and evaluate the situation, is a tool to separate the transactional buyers from the relational buyers. The transactional buyers will be concerned about the diagnostic fee. For them, the amount of the transactional fee doesn't matter. It's the fact that it exists. A transactional buyer will object to the diagnostic fee, whether it's $25, $125, or anywhere in between.

While the diagnostic fee does not have much of a profit margin, it's a good way to determine if you are spending your valuable time with the right client. We talk more about this in Chapter 16. The optimal customers for your HVAC contracting business are relational buyers, not transactional buyers. Relational customers allow you to make a margin so that the business can grow and be around to serve them for years to come.

Flat-Rate Repair Fee

Our industry has various methodologies for pricing repairs, but they all come down to this: repair pricing is the culmination of the hourly cost of doing business for your entire enterprise. As mentioned earlier, we do these as flat-rate pricing, covering all costs.

Replacement Sales

In pricing replacement jobs, you have to serve two masters: gross margin and gross margin per crew day. For replacement pricing, ideally, you will maintain at least a 50 percent gross margin on a job that is completed in one day. If the job to install takes multiple days, make sure you charge the multiple-day labor costs as well as the target gross profit per installation day fees.

I've seen many HVAC contractors become confused because their gross margin is within the 50 percent target, but they are not creating enough gross profit dollars to fund their overhead and profit. This is because they are not charging for the opportunity cost of subsequent installation days.

The best methodology for replacement air conditioning pricing is to calculate a flat-rate price for each of the system types you plan to sell. Create a package that covers the installation in 90 percent of all jobs and train your salespeople to sleuth out the few items from the job survey that aren't typical—that fall into the 10 percent.

Service Agreements

Ideally, you want as many people on service agreements as you possibly can to iron out the seasonality of your business and to help mitigate your lead generation costs. One school of thought in our industry is that we should give maintenance plans away for free. I argue that the customer will not perceive value in the service if it's given away, and it encourages a transactional relationship with your customer. For maintenance pricing, calculate a price that at least covers your cost of maintenance and, ideally, creates a profit. Maintenance plans create long-term relationships with your customers, generate repeat business, and help you maintain your technician base with steady year-round work.

When I coached Leonardo on this pricing formula, he immediately went back to his business and restructured it accordingly. In other words, he paid his technicians and salespeople correctly, so that he could obtain his gross profit per installation, gross margin, and net profit motives.

As soon as he implemented this new model, in the first month, the business made its first significant profit. He continued to be profitable from that point on because he was pricing correctly. After struggling and spending thousands of his hard-earned dollars, hiring one expert after the next, he implemented the advice I gave, and his business flourished. He could move forward.

But I get it. Leonardo was looking for a Prince Charming to save his business, just like I was early on. That made me want to help him help himself. He didn't need a savior; he needed a system. If I hadn't picked up Michael's *E-Myth Revisited*, I would have flailed around and gone out of business, or just lived off my cash flow forever and worked myself to death.

In the next chapter, Michael talks about the lifeblood of our business: *customers*. It's important that we find the right kinds of customers and take care of them, as they will take care of us. ✤

CHAPTER 15

On the Subject
of Customers

Michael E. Gerber

Whether individuals or organizations, we follow those who lead not because we have to, but because we want to. We follow those who lead not for them, but for ourselves.

—Simon Sinek

When it comes to the business of HVAC, the best definition of customers I've ever heard is this:

Customers: *very special people who drive most contractors crazy.*

Does that work for you?

After all, it's a rare customer who shows any appreciation for what an HVAC contractor has to go through to do the job as promised. Don't they always think the price is too high? And don't they focus on problems, broken promises, and the mistakes they think you made, rather than all the ways you bent over backwards to give them what they need?

Do you ever hear other contractors voice these complaints? More to the point, have you ever voiced them yourself? Well, you're not alone. I have yet to meet an HVAC contractor who doesn't suffer from a strong case of customer confusion.

Customer confusion is about:

What your customer really wants

How to communicate effectively with your customer

How to keep your customer truly happy

How to deal with customer dissatisfaction

Whom to call a customer

Confusion 1: What Your Customer Really Wants

Your customers aren't just people; they're very specific kinds of people. Let me share with you the six categories of customers as seen from the E-Myth marketing perspective: (1) tactile customers, (2) neutral customers, (3) withdrawal customers, (4) experimental customers, (5) transitional customers, and (6) traditional customers.

Your entire marketing strategy must be based on which type of customer you are dealing with. Each of the six customer types spends money on HVAC services for very different, and identifiable, reasons. These are:

Tactile customers get their major gratification from interacting with other people.

Neutral customers get their major gratification from interacting with inanimate objects (computers, cars, information).

Withdrawal customers get their major gratification from interacting with ideas (thoughts, concepts, stories).

Experimental customers rationalize their buying decisions by perceiving that what they bought is new, revolutionary, and innovative.

Transitional customers rationalize their buying decisions by perceiving that what they bought is dependable and reliable.

Traditional customers rationalize their buying decisions by perceiving that what they bought is cost-effective, a good deal, and worth the money

In short:

If your customer is tactile, you have to emphasize the *people* of your business.

If your customer is neutral, you have to emphasize the *technology* of your business.

If your customer is a withdrawal customer, you have to emphasize the *idea* of your business.

If your customer is an experimental customer, you have to emphasize the *uniqueness* of your business.

If your customer is transitional, you have to emphasize the *dependability* of your business.

If your customer is traditional, you have to talk about the financial *competitiveness* of your business.

What your customers want is determined by who they are. Who they are is regularly demonstrated by what they do. Think about the customers with whom you do business. Ask yourself: In which of the categories would I place them? What do they do for a living?

If your customer is a mechanical engineer, for example, it's probably safe to assume he's a neutral customer. If another one of your customers is a cardiologist, she's probably tactile. Accountants tend to be traditional, and software engineers are often experimental.

Having an idea about which categories your customers may fall into is very helpful to figuring out what they want. Of course, there's no exact science to it, and human beings constantly defy stereotypes. So don't take my word for it. You'll want to make your own analysis of the customers you serve.

Confusion 2: How to Communicate Effectively with Your Customer

The next step in the customer satisfaction process is to decide how to magnify the characteristics of your business that are most likely to appeal to your preferred category of customer. That begins with what marketing people call your positioning strategy.

What do I mean by positioning your business? You position your business with words. A few well-chosen words to tell your customers exactly what they want to hear. In marketing lingo, those words are called your USP, or unique selling proposition.

For example, if you are targeting tactile customers (ones who love people), your USP could be: "Cozy Home HVAC, where the comfort of people *really* counts!"

If you are targeting experimental customers (ones who love new, revolutionary things), your USP could be: "Smart Home HVAC, where living on the edge is a way of life!" In other words, when they choose to schedule an installation with you, they can count on both your services and equipment to be on the cutting edge of the HVAC industry.

Is this starting to make sense? Do you see how the ordinary things most contractors do to get customers can be done in a significantly more effective way?

Once you understand the essential principles of marketing the E-Myth way, the strategies by which you attract customers can make an enormous difference in your market share.

Confusion 3: How to Keep Your Client Happy

Let's say you've overcome the first three confusions. Great. Now how do you keep your customer happy?

Very simple . . . just keep your promise! And make sure your customer knows you kept your promise every step of the way.

In short, giving your customers what they think they want is the key to keeping your customers (or anyone else, for that matter) really happy.

If your customers need to interact with people (high touch, tactile), make certain that they do.

If they need to interact with things (high tech, neutral), make certain that they do.

If they need to interact with ideas (in their head, withdrawal), make certain that they do.

And so forth.

At E-Myth, we call this your *customer fulfillment system*. It's the step-by-step process by which you do the task you've contracted to do and deliver what you've promised—on time, every time.

But what happens when your customers are not happy? What happens when you've done everything I've mentioned here and your customer is still dissatisfied?

Confusion 4: How to Deal with Customer Dissatisfaction

If you have followed each step up to this point, customer dissatisfaction will be rare. But it can and will still occur—people are people, and some people will always find a way to be dissatisfied with something. Here's what to do about it:

Always listen to what your customers are saying. And never interrupt while they're saying it.

After you're sure you've heard all of your customer's complaint, make absolutely certain you understand what she said by phrasing a question such as: "Can I repeat what you've just told me, Ms. Harton, to make absolutely certain I understand you?"

Secure your customer's acknowledgment that you have heard her complaint accurately.

Apologize for whatever your customer thinks you did that dissatisfied her . . . even if you didn't do it!

After your customer has acknowledged your apology, ask her exactly what would make her happy.

Repeat what your customer told you would make her happy, and get her acknowledgment that you have heard correctly.

If at all possible, give your customer exactly what she has asked for.

You may be thinking, "But what if my customer wants something totally impossible?" Don't worry. If you've followed my recommendations to the letter, what your customer asks for will seldom seem unreasonable.

Confusion 5: Whom to Call a Customer

At this stage, it's important to ask yourself some questions about the kind of customers you hope to attract to your business:

Which types of customers would you most like to do business with?

Where do you see your real market opportunities?

Who would you like to work with, provide services to, and position your business for?

In short, *it's all up to you*. No mystery. No magic. Just a systematic process for shaping your company's future. But you must have the passion to pursue the process. And you must be absolutely clear about every aspect of it.

Until you know your customers as well as you know yourself.

Until all your complaints about customers are a thing of the past.

Until you accept the undeniable fact that customer acquisition and customer satisfaction are more science than art.

But unless you're willing to grow your company, you'd better not follow any of these recommendations. Because if you do what I'm suggesting, it's going to grow.

This brings us to the subject of *growth*. But first, let's see what Ken has to say about *customers*. ♣

CHAPTER
16

The Best Customers

Ken Goodrich

*We see our customers as invited guests to a party, and we are the hosts.
It's our job every day to make every important aspect of the customer
experience a little better.*

—Jeff Bezos

How do you bring in customers? If your customers are anyone other than your mom or favorite uncle, you're going to need a system to bring in sales.

If I wanted my company to have its own proprietary sales system so we could take care of our customers at every touchpoint. To create this system, I hired a notable HVAC sales trainer from outside our company named Pebble. He had all the credentials and experience for the job.

My vision of the sales system was for it to be clear and simple. As I specifically explained, the system would need to be applied to everything we did to help our customers, from the call center to the technicians, as well as our salespeople, of course. I suggested that we use a three- or four-step process and use an acronym to describe each step for ease of memory.

Pebble came in and started working away every day for a year, typing on the computer like crazy. Reams of paper were coming off the printer. I thought he was writing this great book, and I was excited. But weeks went by, along with thousands of dollars for his services.

"Where is the sales system?" I asked him.

"It's called *The P.R.E.C.I.S.E. System*," he said.

"I asked for a four-step process!" I said. "How do you even spell 'precise'?"

"*I* after '*E*' except after '*C*'? I don't know," he answered.

His system required customers to provide utility bills, average rates, and even temperatures each family member liked. He had created a sales system so complicated for both the customers and our team that we couldn't use it.

I was extremely frustrated. I felt like I had wasted a lot of time and money with him, and we parted ways. Sales consultants sometimes end up being more concerned about selling their own services than making something practical for you to get real use out of.

In this chapter, I share our sales systems for identifying the best kinds of customers, then converting and serving those customers. If you have the mindset that your HVAC company is here to help improve your customers' lives, ultimately, those customers will champion your business and help you succeed.

Transactional vs. Relational Customers

Your goal is to create an enterprise that caters to relational customers, as opposed to transactional customers. The transactional customers' only consideration is price. The relational customers' concern is long-term relationships, quality, confidence, and peace of mind. A brilliant marketing strategist, Roy H. Williams, talks about four types of buyers, as shown in this chart.

Time Is Money for Everyone

For your customers, time is money. As business owners, time is money for us, too. We have limited resources in terms of time and employees. We need to direct our resources toward customers of the highest value for our companies.

The customers you want to stay away from are transactional buyers, who are primarily concerned with price. These include those who are (1) "Buried in Obligations," who have no time or money, and (2) "Bargain Hunters," who shop by price comparison.

The highest-quality customers to seek out are relational customers, who appreciate your long-term relationship, quality service, and that you are respectful of their time. These include the (3) "Overworked and Well-Paid," who will pay generously if it buys them time. Make it easy for them. The other type of customers you want are the (4) "Bored, Idle Rich," who have too much time and money. They're looking for an experience or purpose. You want to spend time with these two types of relational customers.

Vetting the Customers in the Call Center

Even if your business isn't large enough for a call center, you can still run your phones with the same system and scale it as you grow. Write scripts for whoever answers the phone to ask key questions that determine which kind of customer you have (transactional or relational) or any potential opportunities technicians should take note of. Here are some talking points:

The diagnostic fee—As I mentioned in Chapter 14, the diagnostic fee is really a tool for assessing your customer. A customer who pushes back on the diagnostic fee may be transactional.

The decision-maker—Another key question would be, "Who is going to be making the decisions today?" If the customer says, "I'm just getting bids and waiting until my spouse gets home," then the call center employee will know not to spend too much time on the call, making a presentation to the wrong person.

The shopaholic—The script can pose a few subtle questions to find out if the customer has had multiple companies bid on the work. If that customer has been dissatisfied with the diagnosis and/or estimates so far, again, that could be a transactional customer.

The equipment—Another way to judge a higher probability for the sale is by having the call center ask the customer, "How old is your air conditioning system?" If the equipment is more than ten years old, the call center flags that customer as an opportunity call. As I mentioned in Chapter 4, the replacement fee is the most lucrative. Knowing the age of the customer's equipment, then, is key.

By setting up certain questions and talking points within the script, anyone who works the phones in your company can ensure that no interaction drains the company's resources unnecessarily, because, remember, time is money. This also frees up resources in the business for relational customers and for filling those revenue buckets.

Respecting and Valuing the Relational Customer

The most important thing to the relational customer is having the relationship with your company. The first step is to create that connection with the customer in order to form the relationship each customer is seeking. We must be very careful with our customers' time and money.

People work hard to support their families. When we come into their homes to repair or replace their air conditioning systems, we must acknowledge the trust they have for us, not only to let us into their homes, but also to part with their hard-earned money so their homes are comfortable for their families.

When you meet with a customer, go in with the problem-solver mindset. Be organized, clean, and professional. Take time to evaluate the situation, and then state your findings accurately and clearly to the customer. This way, you'll have happy customers and certainly more buyers.

Hold in high regard the responsibility that you have to do the job perfectly. Your customers need to feel confident the work will last and that they got more than what they paid for.

We need to respect and protect our customers' property. Do not step your dirty feet on their carpet; put tarps down. Tape plastic against the walls, so you don't ruin their paint. Patch drywall and touch up paint immediately after you're finished. (Include that in the job. Don't make the customer find other contractors to fix what you tore up.) Give them a complete job, from start to finish.

Helping the Customer During Harsh Seasons

During harsh weather—heat or cold—our mantra in the HVAC industry should always be "Leave no customers down." We are in the emergency services business, and you need to structure your business around that.

Go when the customer needs you, not when it's convenient for you. In the high season, we in the HVAC industry know to staff accordingly. Air conditioners do not discriminate; they break every day. You need to have the teams available to serve your customers every single day and not let them down. If you can't get there, call one of your friendly competitors to ask if they'll help you out and take care of your customer for you.

Air conditioning and heating isn't a luxury in many areas; it's a life support system. If it's a warranty service call and there are extreme conditions that cannot be fixed that day, you should consider putting your customer up in a hotel until you get the system fixed. Remember, you need to fill your customer base with relational customers. Your goal is to create and maintain those long-term relationships with these types of customers.

RISE: My Own Four-Step Sales System

After the PRECISE sales system disaster, I decided to create my own system with my entire team. I realized I needed a system that catered to the relational customer. I researched the fundamentals of sales systems and noticed they all had four key things in common:

1. Quickly establish a relationship with the customer.
2. Create a system to inspect the customer's needs.
3. Package solutions to offer the customer.
4. Overcome the customer's objections to purchasing.

From this, we created our own system, and systems within that system, with a simple four-letter acronym: RISE (Relationship, Inspect, Solutions, Execute). RISE is a foundational sales system that's easy for my employees to understand and follow and that customers can appreciate. Here's the breakdown:

1. Create a **relationship** with a customer. We borrowed key concepts from the classic Dale Carnegie book, *How to Win Friends and Influence People*, to train our technicians and salespeople on soft skills, so they know how to be relational with the customers. It helps the technician and salesperson

quickly form a rapport with a customer to establish trust and make the buying process easier.

2. **Inspect** the situation to understand the customer's needs. We created a series of checklists to identify the current conditions of the home construction and HVAC system.

3. Package various **solutions** to provide to the customer. With each of the solutions, performance standards—good, better, best—are listed to fit the customer's needs.

4. **Execute** on closing objections. We wrote scripts for the technicians and salespeople so we can teach them how to overcome typical sales objections, such as "Your price is too high," "I need to talk to my spouse," or "I want to get three bids."

You can develop your own system by thinking in blocks like these. In my system, the front and the back (Relationships, Execute) never change for whatever product or service we are selling. The middle (Inspect, Solutions) changes based on the products or services. As long as you teach your employees how to build a relationship first and overcome closing objections by the customer, you can add the inspection and solutions for every product and service you provide to make the system your own.

Since we've implemented our RISE Sales System, we've established healthy relationships with our customers. They like us, and we like them. We have become the authority in the HVAC industry because we have our system. We have a methodology to inspect their homes; we don't wander around and ask them a bunch of questions they don't know the answers to. We're working with our checklist. Our solutions are already defined.

Everyone on the team is trained and tested on the closing objections to help the customer make a good, informed decision. We have created a very simple system whereby customers can have an enjoyable, consistent experience every time we're in their home. And it's scalable; it grows as we grow.

This brings us into our next topic: *growth*. In the next chapter, Michael has some inspiring words about scaling your business. ❖

On the Subject
of Growth

Michael E. Gerber

All organizations are hierarchical. At each level people serve under those above them. An organization is therefore a structured institution. If it is not structured, it is a mob. Mobs do not get things done, they destroy things.

—Theodore Levitt, *Marketing for Business Growth*

The rule of business growth says that every business, like every child, is destined to grow. Needs to grow. Is determined to grow.

Once you've created your HVAC business, once you've shaped the idea of it, the most natural thing for it to do is to . . . *grow!* And if you stop it from growing, it will die.

Once an HVAC contractor has started a company, it's his or her job to help it grow. To nurture it and support it in every way. To infuse it with

Purpose

Passion

Will

Belief

Personality

Method

As your business grows, it naturally changes. And as it changes from a small business to something much bigger, you will begin to feel out of control. News flash: That's because you *are* out of control.

Your business has exceeded your know-how, sprinted right past you, and now it's taunting you to keep up. That leaves you two choices: Grow as big as your business demands you to grow, or try to hold your business at its present level—at the level you feel most comfortable.

The sad fact is that most contractors do the latter. They try to keep their business small, securely within their comfort zone. Doing what they know how to do, what they feel most comfortable doing. It's called playing it safe.

But as the business grows, the number, scale, and complexity of tasks will grow, too, until they threaten to overwhelm the contractor. More people are needed. More space. More money. Everything seems to be happening at the same time. A hundred balls are in the air at once.

As I've said throughout this book: Most contractors are not entrepreneurs. They aren't true businesspeople at all, but technicians suffering from an entrepreneurial seizure. Their philosophy of coping with the workload can be summarized as "just do it," rather than figuring out how to get it done through other people using innovative systems to produce consistent results.

Given most contractors' inclination to be the master juggler in their business, it's not surprising that as complexity increases, as work expands beyond their ability to do it, as money becomes more elusive, they are just holding on, desperately juggling more and more balls. In the end, most collapse under the strain.

You can't expect your business to stand still. You can't expect your business to stay small. A company that stays small and depends upon you to do everything isn't a company—it's a job!

Yes, just like your children, your business must be allowed to grow, to flourish, to change, to become more than it is. In this way, it will match your vision. And you know all about vision, right? You better. It's what you're thriving to do best!

Do you feel the excitement? You should. After all, you know what your business is but not what it *can be*.

It's either going to grow or die. The choice is yours, but it is a choice that must be made. If you sit back and wait for change to overtake you, you will always have to answer no to this question: Are you ready?

That brings us to the subject of *change*. But first, let's see what Ken has to say about *growth*. ✤

CHAPTER 18

Green and Growing

Ken Goodrich

I believe that if the people who are running and participating in a company grow, then the company's growth in many respects will take care of itself.

—James McNerney

New sales leads can come from anywhere.

In 1990, a colleague who owned a competing HVAC business passed away. He had a well-known business in Las Vegas with many customers. I called the company to see if it was still operating without him, and I got a recording that said, "This phone is no longer in service."

I thought, *I wonder if I can take over his old business phone number?*

I decided to go to the telephone company. At the time, you could take over old numbers by asking the phone company.

When I got to the window, I said, "I believe this telephone number has been disconnected, and I would like to have it."

"Well, we don't do that," said the clerk. "A business phone has to be dormant for one year before we give that number out, in case the old owner wants to come back and claim it."

"I assure you he's not coming back to claim it. He passed away."

"I can't do anything about that, sir," she told me.

I eventually asked to talk to a supervisor. After an hour of waiting, the supervisor came out and said, "If you can prove to us that he's passed away, we'll give you the phone number."

"How am I going to prove to you that he passed away?" I asked.

"I don't know," she said. "Go figure it out."

That's what I did. I bought a copy of the death certificate for $7 and brought it to the telephone company. The supervisor smirked at me, did a few keystrokes on her computer, and said, "The number has been transferred to you."

As soon as I got to the office, my dispatcher told me there was a customer on the line asking if the bid from this other company was still good.

I picked up the telephone and said, "Absolutely, that $7,000 bid is still good."

That $7 certificate eventually turned into millions of dollars for our company.

Your business growth does not have to be an arduous, time-consuming, baby-step process. In this chapter, I talk about how growth, like everything else in your business, is a system. You can strategize three types of growth and scale rapidly with intention. If you do it right, rather than feeling chaotic, you will feel energized and confident. If you have a growth mindset, everyone—from your vendors to your subcontractors, employees, customers, and even partners and investors—will become successful along with you.

Profit and Growth

Someone once said, "Either you're green and growing, or you're brown and dying." We've always used that as an adage around my businesses. I've seen companies failing to grow because they fear the risks associated with growing. By holding themselves back, they fail anyway. I've also seen companies grow for growth's sake, and they too fail. Growth, like all things in a business, needs to have a system.

Every step of the way, you need to be profitable. You can't focus on revenue and forget about profit. When I plan growth, I start at the bottom and work my way up. If I want to produce another million dollars of profit next year, I ask myself, *What are the additional costs I need to incur, in terms of property, plant, equipment, people, or leads, to help me get that extra million?* And then I work backward to put those resources in place.

Organizational Chart for Growth

You need to have an organizational chart that supports and plots the journey of your business' future. When your business is small, the organizational chart might include positions you have to fill yourself. But as you grow, you'll need to put other people in those spots.

Growth requires you to look at your business very objectively. Sometimes the people who helped get your company to a certain level may not have the desire, the ability, or the bandwidth to help take it to the next level. Rather than let those people go, put them on a different seat in the bus. This comes with its own set of challenges, such as hurt feelings or wounded pride at having to take a back seat.

It's a very delicate process. But, delicate or not, you don't want the turnover. Keep the corporate memory in place. As the leader of the enterprise, you have to make some hard decisions for the benefit of the enterprise as a whole. That said, there will be times when you must make the decision to let people move on from the company to benefit the enterprise. These are skills you will sharpen as your company grows.

Stay Out Front

You have an obligation to your people and your enterprise to stay out front. You stay out front by admitting that you don't have all the answers all the time. Leadership is about having the ability and humility to go out and hire people who are better, smarter, and faster than you without being intimidated or fearful.

I've made big leaps in my business by hiring people with Ivy League degrees and much more business experience than I have. But I stay out front by not being afraid of them and letting them take the ball and run with it while I provide the vision and support.

Meanwhile, I keep refining my own skills. Every owner of a business needs to be continually sharpening his or her own saw. I have read something every single day, for thirty-two years, on the topic of business in general or the HVAC business. I might have read for a minute on some days, and two hours on others. It's an easy habit to maintain because I have a smartphone and can visit sites such as serviceroundtable.com. I attribute a lot of my personal growth to sharpening my saw every day.

I've also been very involved in the industry by participating in local and national associations and best practice groups. While some groups include my local competitors, it is far more important to learn from one another and share methodologies. I attribute a lot of my success to making a lifelong commitment to the industry.

All of these things help me grow, which in turn helps the operation grow. Knowledge, education, and experience give me a much better perspective. Because, as the best of the best say, "You don't know what you don't know."

Calculating Your Place in the Market and Your Potential

As you're building your business, rather than planning to grow at a typical rate of 10 percent or so a year, I again challenge you to define what your true potential is. Do not tether yourself to last year or last month or even last week. You need to pull yourself back from the day-to-day and watch the game being played in order to identify places where you can make great growth gains.

What is the true potential of the marketplace you're serving today? In our particular business, that's pretty easy to define, because we can get those figures from the American Heating Refrigeration Institute (AHRI). AHRI will tell you exactly how many air conditioners will be shipped to your market in the next year for replacement sales.

By multiplying that number of units by what you deem to be the average sale price, you can determine the value of your market. Most major US markets are in the billions of dollars. If you ever want to see what your place is in the world, and therefore your true potential, take your revenue and divide it by that billion-dollar figure. That'll tell you what market share you really have.

Evaluate Your Weaknesses

Once you see what your market is worth, for example, $1.6 billion, you have a whole other perspective on how to grow your business. Rather than growing your business according to your own limitations, your business systems, or your number of employees, it's incumbent upon you as the business owner to recognize where your weak spots are and fix them. This is how you can grab more market share, more growth, and corresponding profits.

If you see that you need to upgrade your business systems, train your employees to give them greater skills, or hire more staff, do it right away. It's the only way you can grow your service into a business, or your business into an enterprise.

To Expand or Not to Expand

The purpose of adding another service in order to grow is to leverage your existing customer base by offering more products or services. It's the nature of most contractors: as soon as they get a reasonable momentum in their HVAC businesses, they want to expand by adding another service, such as plumbing or electrical. I see many contractors put their business in a downward spiral, trying to add another trade too soon. If you build your HVAC business correctly, with a solid foundation, with all of your systems in place, your business will continue to grow, while you focus on your core competency: HVAC.

But, once you get your business and the team growing at a commendable, consistent rate, for example, 20 percent or more each year, with your processes and systems and management team in place, and you're consistently expanding, then and only then do you add another trade.

Three Types of Acquisitions

Growth by acquisitions is a powerful way to expand. It is never too early to expand your customer base. I made my first acquisition when I was still in a service van. Throughout my career, I have completed 107 acquisitions and am still counting. There are three types of acquisitions: phone number/database, tuck-in, and strategic.

1. Phone number/database—This is what I did with the death certificate. With a phone number/database acquisition, you gain customers by buying the trade name, telephone number, database, and URL of a small or failing company. With this type of acquisition, you are essentially only buying leads. If done correctly, you can purchase for one third less than traditional lead generation costs.

2. Tuck-in—A tuck-in acquisition is where you'll buy a company with some assets. You may inherit some personnel

and loyal customers. You may try to keep the trade name alive, for a period of time, until those customers have become familiar with your company. In general, the trade name for the company you acquire gets "tucked in" to yours (it goes away). This is typically a smaller deal than what is involved in a strategic acquisition.

3. Strategic—Strategic acquisition is the largest acquisition of the three types. In a strategic acquisition, you buy a company similar to your current size or larger, and you absorb the trade name completely into your company, merge the trade names together, or operate it separately. Either way, try to keep the company you acquired together in terms of property, plant, equipment, and people. At one point, I ran three different entities in the same market. All three companies used my same systems, but because they had different brand names in different locations, they attracted different customers.

Open your eyes to the many opportunities that will come along to grow your company. For a $7 investment and a little bit of creativity on my part, I added $500,000 in revenue that year to my business from one phone number. If you have a growth mindset, you'll be bold and unafraid to try new things.

This leads us to our next chapter, where Michael talks about change and the idea of *expansion* and *contraction*. ❧

On the Subject
of Change

Michael E. Gerber

Our Leaders of today need the philosophy of the past, paired with the scientific knowledge and technology of tomorrow.

—Anders Indset

So your company is growing. That means, of course, that it's also changing. Which means it's driving you and everyone in your life crazy.

That's because, to most people, change is a diabolical thing. Tell most people they've got to change, and their first instinct is to crawl into a hole. Nothing threatens their existence more than change. Nothing cements their resistance more than change. Nothing.

Yet for the past forty years, that's exactly what I've been proposing to small business owners: the need to change. Not for the sake of change itself, but for the sake of their lives.

I've talked to countless contractors whose hopes weren't being realized through their business; whose lives were consumed by work;

135

who slaved increasingly longer hours for decreasing pay; whose dissatisfaction grew as their enjoyment shriveled; whose business had become the worst job in the world; whose money was out of control; whose employees were a source of never-ending hassles, just like their customers, their bank, and, increasingly, even their family.

More and more, these contractors spent their time alone, dreading the unknown and anxious about the future. And even when they were with people, they didn't know how to relax. Their mind was always on the job. They were distracted by work, by the thought of work. By the fear of falling behind.

And yet, when confronted with their condition and offered an alternative, most of the same contractors strenuously resisted it. They assumed that if there were a better way of doing business, they already would have figured it out. They derived comfort from knowing what they believed they already knew. They accepted the limitations of being an HVAC contractor; or the truth about people; or the limitations of what they could expect from their customers, their employees, their subcontractors, their bankers—even their family and friends.

In short, most contractors I've met over the years would rather live with the frustrations they already have than risk enduring new frustrations.

Isn't that true of most people you know? Rather than opening up to the infinite number of possibilities life offers, they prefer to shut their lives down to respectable limits. After all, isn't that the most reasonable way to live?

I think not. I think we must learn to let go. I think that if you fail to embrace change, it will inevitably destroy you.

Conversely, by opening yourself to change, you give your HVAC business the opportunity to get the most from your talents.

Let me share with you an original way to think about change, about life, about who we are and what we do. About the stunning notion of expansion and contraction.

Contraction versus Expansion

"Our salvation," a wise man once said, "is to allow." That is, to be open, to let go of our beliefs, to change. Only then can we move from a point of view to a viewing point.

That wise man was Thaddeus Golas, the author of a small, powerful book entitled *The Lazy Man's Guide to Enlightenment* (Seed Center, 1971).

Among the many inspirational things he had to say was this compelling idea:

The basic function of each being is expanding and contracting. Expanded beings are permeable; contracted beings are dense and impermeable. Therefore each of us, alone or in combination, may appear as space, energy, or mass, depending on the ratio of expansion to contraction chosen, and what kind of vibrations each of us expresses by alternating expansion and contraction. Each being controls his own vibrations.

In other words, Golas tells us that the entire mystery of life can be summed up in two words: *expansion* and *contraction*. He goes on to say:

We experience expansion as awareness, comprehension, understanding, or whatever we wish to call it.

When we are completely expanded, we have a feeling of total awareness, of being one with all life.

At that level we have no resistance to any vibrations or interactions of other beings. It is timeless bliss, with unlimited choice of consciousness, perception, and feeling.

When a [human] being is totally contracted, he is a mass particle, completely imploded.

To the degree that he is contracted, a being is unable to be in the same space with others, so contraction is felt as fear, pain, unconsciousness, ignorance, hatred, evil, and a whole host of strange feelings.

At an extreme [of contraction, a human being] has the feeling of being completely insane, of resisting everyone and everything, of being unable to choose the content of his consciousness.

Of course, these are just the feelings appropriate to mass vibration levels, and he can get out of them at any time by expanding, by letting go of all resistance to what he thinks, sees, or feels.

Stay with me here. Because what Golas says is profoundly important. When you're feeling oppressed, overwhelmed, exhausted by more than you can control—contracted, as Golas puts it—you can change your state to one of expansion.

According to Golas, the more contracted we are, the more threatened by change; the more expanded we are, the more open to change.

In our most enlightened—that is, open—state, change is as welcome as non-change. Everything is perceived as a part of ourselves. There is no inside or outside. Everything is one thing. Our sense of isolation is transformed to a feeling of ease, of light, of joyful relationship with everything.

As infants, we didn't even think of change in the same way, because we lived those first days in an unthreatened state. Insensitive to the threat of loss, most young children are only aware of *what is*. Change is simply another form of *what is*. Change just *is*.

However, when we are in our most contracted—that is, closed—state, change is the most extreme threat. If the known is what I have, then the unknown must be what threatens to take away what I have. Change, then, is the unknown. And the unknown is fear. It's like being between trapezes.

To the fearful, change is threatening because things may get worse.

To the hopeful, change is encouraging because things may get better.

To the confident, change is inspiring because the challenge exists to improve things.

If you are fearful, you see difficulties in every opportunity. If you are fear-free, you see opportunities in every difficulty.

Fear protects what I have from being taken away. But it also disconnects me from the rest of the world. In other words, fear keeps me separate and alone.

Here's the exciting part of Golas' message: with this new understanding of contraction and expansion, we can become completely attuned to where we are at all times.

If I am afraid, suspicious, skeptical, and resistant, I am in a contracted state. If I am joyful, open, interested, and willing, I am in an expanded state. Just knowing this puts me on an expanded path. Always remembering this, Golas says, brings enlightenment, which opens me even more.

Such openness gives me the ability to freely access my options. And taking advantage of options is the best part of change. Just as there are infinite ways to greet a client, there are infinite ways to run your company. If you believe Thaddeus Golas, your most exciting option is to be open to all of them.

Because your life is lived on a continuum between the most contracted and the most expanded—the most closed and most open—states, change is best understood as the movement from one to the other, and back again.

Most small business owners I've met see change as a thing in itself, as something that just happens to them. Most experience change as a threat. Whenever change shows up at the door, they quickly slam it. Many bolt the door and pile up the furniture. Some even run for their gun.

Few of them understand that change isn't a thing in itself, but rather the manifestation of many things. You might call it the revelation of all possibilities. Think of it as the ability at any moment to sacrifice what we are for what we could become.

Change can either challenge us or threaten us. It's our choice. Our attitude toward change can either pave the way to success or throw up a roadblock.

Change is where opportunity lives. Without change we would stay exactly as we are. The universe would be frozen still. Time would end.

At any given moment, we are somewhere on the path between a contracted and expanded state. Most of us are in the middle of the journey, neither totally closed nor totally open. According to

Golas, change is our movement from one place in the middle toward one of the two ends.

Do you want to move toward contraction or toward enlightenment? Because without change, you are hopelessly stuck with what you've got.

Without change,

we have no hope;

we cannot know true joy;

we will not get better; and

we will continue to focus exclusively on what we have and the threat of losing it.

All of this negativity contracts us even more, until, at the extreme closed end of the spectrum, we become a black hole so dense that no light can escape.

Sadly, the harder we try to hold on to what we've got, the less able we are to do so. So we try still harder, which eventually drags us even deeper into the black hole of contraction.

Are you like that? Do you know anybody who is?

Think of change as the movement between where we are and where we're not. That leaves only two directions for change: either moving forward or slipping backward. We either become more contracted or more expanded.

The next step is to link change to how we feel. If we feel afraid, change is dragging us backward. If we feel open, change is pushing us forward.

Change is not a thing in itself, but a movement of our consciousness. By tuning in, by paying attention, we get clues to the state of our being.

Change, then, is not an outcome or something to be acquired. Change is a shift of our consciousness, of our being, of our humanity, of our attention, of our relationship with all other beings in the universe.

We are either "more in relationship" or "less in relationship." Change is the movement in either of those directions. The exciting part is that *we possess the ability to decide which way we go . . . and to know in the moment which way we're moving.*

Closed, open . . . Open, closed. Two directions in the universe. The choice is yours.

Do you see the profound opportunity available to you? What an extraordinary way to live!

Enlightenment is not reserved for the sainted. Rather, it comes to us as we become more sensitive to ourselves. Eventually, we become our own guides, alerting ourselves to our state, moment by moment: *open . . . closed . . . open . . . closed.*

Listen to your inner voice, your ally, and feel what it's like to be open and closed. Experience the instant of choice in both directions.

You will feel the awareness growing. It may be only a flash at first, so be alert. This feeling is accessible, but only if you avoid the black hole of contraction.

Are you afraid that you're totally contracted? Don't be—it's doubtful. The fact that you're still reading this book suggests that you're moving in the opposite direction.

You're more like a running back seeking the open field. You can see the opportunity gleaming in the distance. In the open direction.

Understand that I'm not saying that change itself is a point on the path; rather, it's the all-important movement.

Change is *in you*, not *out there*.

What path are you on? The path of liberation? Or the path of crystallization?

As we know, change can be for the better or for the worse.

If change is happening *inside* of you, it is for the worse only if you remain closed to it. The key, then, is your attitude—your acceptance or rejection of change. Change can be for the better only if you accept it. And it will certainly be for the worse if you don't.

Remember, change is nothing in itself. Without you, change doesn't exist. Change is happening inside of each of us, giving us clues to where we are at any point in time.

Rejoice in change, for it's a sign you are alive.

Are we open? Are we closed? If we're open, good things are bound to happen. If we're closed, things will only get worse.

According to Golas, it's as simple as that. Whatever happens defines where we are. *How* we are is *where* we are. It cannot be any other way.

For change is life.

Charles Darwin wrote, "It is not the strongest of the species that survive, nor the most intelligent, but the one that proves itself most responsive to change."

The growth of your HVAC business, then, is its change. Your role is to go with it, to be with it, to share the joy, embrace the opportunities, meet the challenges, learn the lessons.

Remember, there are three kinds of people: (1) those who make things happen, (2) those who let things happen, and (3) those who wonder what the hell happened. The people who make things happen are masters of change. The other two are its victims.

Which type are you?

The Big Change

If all this is going to mean anything to the life of your company, you have to know when you're going to leave it. At what point, in your company's rise from where it is now to where it can ultimately grow, are you going to sell it? Because if you don't have a clear picture of when you want out, your business is the master of your destiny, not the reverse.

As we stated earlier, the most valuable form of money is equity, and unless your business vision includes your equity and how you will use it to your advantage, you will forever be consumed by your business.

Your business is potentially the best friend you ever had. It is your business's nature to serve you, so let it. If, however, you are not a wise steward, if you do not tell your business what you expect from it, it will run rampant, abuse you, use you, and confuse you.

Change. Growth. Equity.

Focus on the point in the future when you will take leave of your business. Now reconsider your goals in that context. Be specific. Write them down.

Skipping this step is like tiptoeing through earthquake country. Who can say where the fault line is waiting? And who knows exactly when your whole world may come crashing down around you?

Which brings us to the subject of *time*. But first, let's see what Ken has to say regarding *change*. ✿

CHAPTER 20

The Constancy of Change

Ken Goodrich

Sometimes if you want to see change for the better, you have to take things into your own hands.

—Clint Eastwood

I n the early 1990s, I was involved in just about everything an HVAC contractor can do in the trade: residential, commercial, sheet metal, kitchens, and industrial work. I was doing everything I could get my hands on. This made me a jack of all trades, master of none. My company prided itself on being able to do everything. But we were still struggling and weren't making much money.

I made a conscious decision to change. The most profitable business we had was the residential service and replacement business. We decided to focus our efforts on that side of the business. I sold off my sheet metal equipment, and my commercial business. I divested myself of all of that stuff.

One day, I happened to be in the office of the man I had sold it all to. He had a family-run company, so his wife and kids were there with him. He said, in a lighthearted way, "The only thing consistent about Ken Goodrich is change." They all laughed.

Sadly, after several years of struggling, he went out of business, while my business continued to thrive.

In this chapter, I talk about how change comes from the top down. I also walk you through the process of selling your business—or the final change as you work your way to financial freedom and your ultimate goal.

Your ability to quickly change your course—whether it's diversifying your company or pulling back to focus on one thing—will lead you to great success. Having an organization that is successful with change also helps employees, vendors, and investors not to fear change, but to understand the necessity and the power of change.

Change and the E-Myth Way

My philosophy on change is that once you understand the E-Myth systems approach to business, change isn't a scary or risky thing anymore. If you're doing one thing and suddenly realize that to keep up with changing conditions in your industry, you need to change, it's simple. Just modify the systems and train your employees. Change is more about you understanding that your business is just a set of systems. You can modify the systems as much as you need to in order to maximize sales, profits, and opportunities while mitigating risk.

The Need for Change

During one of my early acquisitions, I bought a fifty-three-year-old company, in which the owner and three other technicians did all the work. When we toured the facility, the owner took me back into his warehouse. The space was full of inventory that was more than twenty years old. He proudly told me, "If anyone ever needs their twenty-year-old air conditioner fixed, I'll be the only guy in town who has the parts."

I thought, iIf anybody has a broken-down, twenty-year-old air conditioner, what they really need is a brand-new air conditioner, especially with all the advanced technology out there. But I just smiled and nodded to the owner when he showed me his warehouse stockpile.

Next up on the tour, he showed me his fleet of old mini panel vans. They looked like old-fashioned milk trucks. He fondly maintained them because they were fuel-efficient and a trademark of the company.

When we reached the front of his dilapidated building, the owner said, "You've got to keep this location, because everyone knows about it. Customers always tell me that their eyes perk up when they drive by our shop."

The stockpile of twenty-year-old parts, the fleet of old milk trucks, and the dilapidated building were sources of pride for the owner. If I had told him these things needed to be changed, he would have taken it personally.

The day after we closed the transaction, two industrial trash bins got loaded into the yard. I had my crew throw away the entire warehouse full of dusty, obsolete parts. A car carrier pulled up, dropped off six brand-new Chevy vans, and hauled off the milk trucks for trade-in. Shortly after, movers showed up and grabbed all the office files we wanted to keep. We were ready to move into the brand-new facility we rented across town.

I ended up discarding these prized parts of the previous owner's business because the actual value of the business was in the customer database and the relationships he had built with those customers. Twelve months after our purchase, we grew the business ten times over. The previous owner was reluctant to change, while I was not, and he missed out on all that growth.

The HVAC industry is changing at a rapid rate. With the rise of e-Commerce, we will see this industry turn over. In the next few years, it is going to be completely different. HVAC contractors who don't understand the E-Myth systems approach will become labor brokers, installing air conditioners for e-Commerce giants such as Amazon, Home Depot, or Walmart.

Our industry is also innovating. With the explosion of the Internet of Things (IoT), smart homes have devices everywhere talking to each other. Every new air conditioner can be remotely controlled by Wi-Fi thermostat and smartphone app.

These new tech gadgets force you, the HVAC contractor, to make a choice: either capitalize on the changes by modifying your systems or do nothing and let innovation put you out of business. Embrace the fact that business (and life) is all about change. And change isn't scary.

Anticipating Change

I mentioned in Chapter 18 the need to constantly educate yourself. In the case of change, if you don't know what's heading your way, you're going to be blindsided. Contractors who fear or are in denial of change see opportunities as threats.

So be smart. Read trade magazines. Keep your eyes and ears open. Rather than complain about changes, use them to grow your business. I can't emphasize enough how important it is to stay alert. Don't sit on the rocks while the waves crash on you every couple of minutes. Systematize your business and be ready for the tides to turn.

The Selling to Grow Mindset

In Chapter 4, I wrote about how you want to run your business as if you were trying to sell it, even if you have no intention of doing so. If you build your company with the mindset that you're going to sell it one day, you'll have a well-organized business, run by a management team that is highly profitable and growing, which is the best and most valuable business you can have.

One idea I embraced after reading the E-Myth book back in 1988 was that the business is not your life. It's something you own for the purpose of funding your primary aim in life. My business is no different than owning a piece of rental property. I maintain it, so that it will get the highest and best rents available, and once the market is right, I sell it for the highest value and move forward.

Making Yourself Attractive to Buyers

Selling your company can be the biggest change of all. For some, it's the ultimate goal, freeing you up to pursue your next aim in life, or perhaps a new business. If you're going to build a successful company to sell with a multiple of its earnings, here are some value building blocks your company should attain to be attractive to buyers:

1. A respectable amount of revenue and profits—Because the businesses in the HVAC industry are typically valued on a multiple of profits, the more profits, the more value for the business.
2. Revenue and scale—A strategic buyer wants a minimum of $10 million in revenue. A $20 million revenue business raises even more eyebrows in interest.
3. Proven systems and a strong management team.
4. Predictable revenue and profit growth.
5. At least two-thirds of your service calls are for customer service agreements.

You can still be the CEO, as most likely you'll stay with the business for a period of time for an easy handoff. Your business is more valuable if you can demonstrate it is not run by you, but by a management team, and is operating with a full set of proven business systems.

How to Sell When You're Ready

Once you attain the value building blocks, you can market the company to potential buyers. I suggest you partner with an investment banking firm and not a business broker. An investment banking firm will help package your business for sale at its highest value. These kinds of firms achieve this by helping to build the story of your business. They also help you make changes while you're looking for a buyer to increase your company's value.

Marketing to Buyers

An investment banking firm will put the presentations together by creating a book for your company. A book contains all the deep pertinent details of your business so that buyers can make informed offers.

You will take the book with you as you visit the potential buyers with your investment banking firm. At the pitches, you'll talk about your business and introduce your management team. The firm will shop your business to the investment community, the strategic buyers, to get the highest and best offer for the company.

Bids and Letters of Intent

Soon, you'll start receiving offers in the form of non-binding letters of intent. You can decide which offer makes the most sense for you and yields you the best return. Your investment banking firm will also share with you which of the companies have the highest probability of closing the transaction. When we're talking millions of dollars, you need to make sure that you're talking with qualified buyers.

After you accept the letter of intent, or negotiate some finer terms for the offer, and get a revised letter of intent, then you'll enter into a purchase sale agreement (PSA). The PSA outlines the terms of the sale. It'll specify the sale price, how you're going to be paid, and representations and warranties that you give to the buyers to make sure they understand that you are delivering the assets in legal forms without encumbrance, or future encumbrances.

Change can be frightening, but the key is to be prepared. With the right mindset, you can lead your organization, your people, and your industry into change for the better.

With all these systems and changes that you need to work on, it's easy to feel overwhelmed. Take a breath. In the next chapter, Michael shares how to free up some *time* to get working on your business. ✤

On the Subject
of Time

Michael E. Gerber

Money and time are the heaviest burdens of life, and the unhappiest of all mortals are those who have more of either than they know how to use.

—Samuel Johnson

"I'm running out of time!" contractors often lament. "I've got to learn how to manage my time more carefully!"

Of course, they see no real solution to this problem. They're just worrying the subject to death. Singing the contractor's blues.

Some make a real effort to control time. Maybe they go to time management classes, or faithfully try to record their activities during every hour of the day.

But it's hopeless. Even when contractors work harder, even when they keep precise records of their time, there's always a shortage of it. It's as if they're looking at a square clock in a round universe. Something doesn't fit. The result: the contractor is constantly chasing work, money, life.

And the reason is simple. Contractors don't see time for what it really is. They think of time with a small "t," rather than Time with a capital "T."

Yet, Time is simply another word for *your life*. It's your ultimate asset, your gift at birth—and you can spend it any way you want. Do you know how you want to spend it? Do you have a plan?

How do *you* deal with Time? Are you even conscious of it? If you are, I bet you are constantly locked into either the future or the past. Relying on either memory or imagination.

Do you recognize these voices? "Once I get through this, I can have a drink . . . go on a vacation . . . retire." "I remember when I was young and installing HVAC was satisfying."

As you go to bed at midnight, are you thinking about waking up at 5:00 a.m. so that you can get to the office by 6:00 a.m. so that you can get to the jobsite by 7:00 a.m., and you've got a full schedule and a new customer scheduled for 2:30?

Most of us are prisoners of the future or the past. While pinballing between the two, we miss the richest moments of our life—the present. Trapped forever in memory or imagination, we are strangers to the here and now. Our future is nothing more than an extension of our past, and the present is merely the background.

It's sobering to think that right now each of us is at a precise spot somewhere between the beginning of our Time (our birth) and the end of our Time (our death).

No wonder everyone frets about Time. What really terrifies us is that *we're using up our life and we can't stop it.*

It feels as if we're plummeting toward the end with nothing to break our free fall. Time is out of control! Understandably, this is horrifying, mostly because the real issue is not time with a small "t" but Death with a big "D."

From the depths of our existential anxiety, we try to put Time in a different perspective—all the while pretending we can manage it. We talk about Time as though it were something other than what it is. "Time is money," we announce, as though that explains it.

But what every contractor should know is that Time is life. And Time ends! Life ends!

The big, walloping, irresolvable problem is that *we don't know how much Time we have left.*

Do you feel the fear? Do you want to get over it?

Let's look at Time more seriously.

To fully grasp Time with a capital "T," you have to ask the Big Question: *How do I wish to spend the rest of my Time?*

Because I can assure you that if you don't ask that Big Question with a big "Q," you will forever be assailed by the little questions. You'll shrink the whole of your life to *this time* and *next time* and the *last time*—all the while wondering, *what time is it?*

It's like running around the deck of a sinking ship worrying about where you left the keys to your cabin.

You must accept that you have only so much Time; that you're using up that Time second by precious second. And that your Time, your life, is the most valuable asset you have. Of course, you can use your Time any way you want. But unless you choose to use it as richly, as rewardingly, as excitingly, as intelligently, as *intentionally* as possible, you'll squander it and fail to appreciate it.

Indeed, if you are oblivious to the value of your Time, you'll commit the single greatest sin: You will live your life unconscious of its passing you by.

Until you deal with Time with a capital "T," you'll worry about time with a small "t" until you have no Time—or life—left. Then your Time will be history . . . along with your life.

I can anticipate the question: If Time is the problem, why not just take on fewer customers? Well, that's certainly an option, but probably not necessary. I know an HVAC contractor with a business that sees three times as many customers as the average, yet he doesn't work long hours. How is it possible?

This contractor has a system. Roughly 50 percent of what needs to be communicated to customers is "downloaded" to the technicians and office staff. By using this expert system, the employees can do everything the contractor or his subcontractors would do—everything that isn't contractor-dependent.

Be Versus Do

Remember when we all asked, "What do I want to be when I grow up?" It was one of our biggest concerns as children.

Notice that the question isn't, "What do I want to *do* when I grow up?" It's "What do I want to *be?*"

Shakespeare wrote, "To be or not to be." Not "To do or not to do."

But when you grow up, people always ask you, "What do you *do?*" How did the question change from *being* to *doing?* How did we miss the critical distinction between the two?

Even as children, we sensed the distinction. The real question we were asking was not what we would end up *doing* when we grew up, but who we would *be.*

We were talking about a *life* choice, not a *work* choice. We instinctively saw it as a matter of how we spend our Time, not what we do *in* time.

Look to children for guidance. I believGood e that as children we instinctively saw Time as life and tried to use it wisely. As children, we wanted to make a life choice, not a work choice. As children, we didn't know—or care—that work had to be done on time, on budget.

Until you see Time for what it really is—your life span—you will always ask the wrong question.

Until you embrace the whole of your Time and shape it accordingly, you will never be able to fully appreciate the moment.

Until you fully appreciate every second that comprises Time, you will never be sufficiently motivated to live those seconds fully.

Until you're sufficiently motivated to live those seconds fully, you will never see fit to change the way you are. You will never take the quality and sanctity of Time seriously.

And unless you take the sanctity of Time seriously, you will continue to struggle to catch up with something behind you. Your frustrations will mount as you try to snatch the second that just whisked by.

If you constantly fret about time with a small "t," then big-"T" Time will blow right past you. And you'll miss the whole point, the real truth about Time: You can't manage it; you never could. You can only live it.

And so that leaves you with these questions: How do I live my life? How do I give significance to it? How can I be here now, in this moment?

Once you begin to ask these questions, you'll find yourself moving toward a much fuller, richer life. But if you continue to be caught up in the banal work you do every day, you're never going to find the time to take a deep breath, exhale, and be present in the now.

So let's talk about the subject of *work*. But first, let's read what Ken has to say about *time*. ✤

CHAPTER 22

Own Your Time

Ken Goodrich

I have two kinds of problems, the urgent and the important. The urgent are not important, and the important are never urgent.

—Dwight D. Eisenhower

After I read *The E-Myth Revisited™*, I finally understood my true role in the company. I was not a technician or manager, but an entrepreneur.

To remind myself of this role, I put a mirror inside my office, adjacent to the door. It was a simple, rectangular sixteen-by-twenty-inch frameless mirror. Every evening when I left the office, I looked myself in the mirror and asked this question: *Did you accomplish something today or did you just prolong the agony?*

The agony, as Michael E. Gerber says, was *doing it, doing it, doing it*—grinding away with no end in sight. Accomplishing meant doing at least one thing the E-Myth way—documenting and implementing a system in order to move forward.

159

At the time, I worked two jobs: technician and CEO. The problem was, I was both a full-time technician and a part-time CEO putting out fires. Once I got *The E-Myth* book, I became keenly aware that to get through the fire, I needed to quantify my systems. But I was stuck. I didn't have the time to work on my business, rather than in my business.

Every day, when I looked in that frameless mirror, I asked myself that burning question. I did that for days, months, years—until I was no longer "doing it" every day. By then, I had grown into the entrepreneur's role, and my time was spent innovating, implementing, and documenting systems.

That's how I sold my first three companies. After that, I had the knowledge and systems to start the next company right. By then, I had developed the systems for managing my time.

In this chapter, I show you how to free up your time for yourself and your employees. When you take the time to get organized, your company will run like a well-oiled machine, giving you even more time.

No one should be working late into the night every night fixing air conditioners. Your business systems should center on efficiency and respect for everyone's time in the organization. When you do that, you and your people will have well-rounded lives, with time for your families and yourselves. Your own time will be focused on *accomplishing* your goals and enjoying your life.

Your Role as an Entrepreneur

As the entrepreneur/CEO of your own HVAC company, you are always in complete control over your time. Every moment you dedicate to a task has a direct impact on your organization. It's very important you manage your time with your plans and goals in mind. If it doesn't move your business forward, it's not worth your time. Your role as a leader is to manage your time, your people's time, and your customers' time. You need to set up systems that best use those three types of time.

Your time is never well used by doing the work. That's not your role. Your role is to be the leader and recruit quality people to accomplish your mission and vision for your company. Your goal is to maximize shareholder value, especially if you're the only shareholder. You can do this in a million ways, but here are the key ones to focus on:

1. Innovate and document the process.
2. Recruit the best and brightest people.
3. Seek acquisitions/growth opportunities.
4. Stay on top of industry trends.
5. Develop your people.

The more refined systems you have, the more time you have. Spend your time developing and refining your systems.

Running an HVAC business is like building a finely tuned machine. Stand back and look at it, instead of keeping your head stuck inside the machine's inner workings. Your job isn't to turn the cogs or fire the pistons. You are the inventor and architect of the machine that is your HVAC business. Focus your time on building the machine.

I work on the machine during my "season" from October to April and then watch my systems in play during the HVAC business peak season from May to September. But I still demand twelve-month profitability in my HVAC business.

How I Rolled Up My Sleeves

Early on, I realized I needed to focus my time on the highest return for the enterprise. I defined how much cash flow I needed—the bare minimum—to keep my current situation steady and pay myself a reasonable living. I also drove our systems and people to make sure we reached that every day. This freed up the capital to make the best decision I could at that point. As I described in Chapter 10, I hired a retired military officer, Curtis, as my first manager. He took the daily work off my shoulders, so I had time to do the work needed to move the organization forward.

Peaks and Shoulders

As you know, the peak season for the HVAC industry is during the summertime. In HVAC, you never have enough people or hours in the day during the summer to complete all the work that comes your way.

By having a clear method to discern which tasks are most important, you can complete the most valuable work, rather than all of the work that gets thrown at you. Quantify the most important activities of the business. Measure them daily. Then, let those measurements tell you what to do every day and exactly how to manage your time.

President Dwight D. Eisenhower built a methodology for making sure that his time was spent productively.

	URGENT	NON-URGENT
IMPORTANT	**DO** Important and Urgent	**SCHEDULE** Important but Not Urgent
NOT IMPORTANT	**DELEGATE** Not Important but Urgent	**ELIMINATE** Not Important and Not Urgent

To figure out where a task fits within this matrix, simply ask yourself two questions: *Is this urgent?* and *Is this important?* As the leader, your focus should be exclusively on the tasks that are important but not urgent. If a task is both urgent and important, it's a crisis that a manager can handle, immediately. The system delegates tasks that are urgent but not important and not urgent and unimportant.

Prioritizing Your Time

The HVAC industry typically has one big season and two shoulder seasons where we see the most demand. This is where the Eisenhower system of time management is a critical part of maximizing your revenue. If you plan ahead and create a system of classifying the demand of work during these busy times into the four quadrants, you will be able to maximize productivity in the face of what seems like a nearly insurmountable quantity of work.

Here is how it works. In the middle of a heatwave, your call center gets a call to do a minor tune-up on a customer's air conditioning. At the same time, another call comes in from a desperate customer with no air conditioning at all. Your call center should be trained to vet the calls by looking at the current volume and weighing that information with weather forecasts. They should be able to pick the times that are ideal for pursuing those less urgent calls.

Remember, we have a limited resource in our industry: skilled labor. The entire HVAC industry is challenged with finding enough quality people. Therefore, your job as the leader is to make sure that your systems and processes maximize the revenues and profits of this constrained resource.

The Time System in Action

The majority of my leadership function in the business today I can do in a few hours. Each week, I have two meetings that last up to one-and-a-half hours, and every month, I have one meeting that goes for six hours.

These days, I purposely don't go to the office before 10:00 a.m. I focus my time in the morning on enjoying my family, exercising, making key phone calls, and reading important emails or projects. I use my workday to observe what's going on. I look at the metrics of our business systems while they're happening, to see if I can pick up any ideas to help them improve. I make notes about any opportunities

I may see. I try not to burden my day with superfluous meetings. I stay fluid. My time is spent thinking about and internalizing the business in order to further my vision.

I see no merit in showing up at 6:00 a.m. and leaving at 6:00 p.m. every night. I've never seen anyone who has had any significant success due to this kind of schedule. Anyone doing that is working in the business instead of on it. Because of Michael's teachings, I decided early on that my business is not *me*. It is something that I own that will help me achieve my primary aims in life.

Because I am not my business, I don't have to be there on a day-to-day basis for it to run smoothly. Being a workaholic *does not* equate to having a successful HVAC business. In my world, I have a beach house where I can spend most of the summer, while my business is running efficiently and effectively right in the heart of the desert.

These days, when I look in the mirror, I still ask myself: *Did you accomplish something today or just prolong the agony?*

But my answer is always about accomplishment. Never agony.

When you focus your time, you will be able to do the same. You will be standing back, instead of clomping through the day-to-day weeds of your business. You will be optimizing your time, making sure every aspect of the business runs with the greatest productivity and efficiency. You will be creating the revenue and profit that fit your ultimate plan.

Next, let's learn what Michael has to say about strategic *work* versus tactical *work*. ✤

CHAPTER 23

On the Subject
of Work

Michael E. Gerber

They intoxicate themselves with work so they won't see how they really are.

—Aldous Huxley

In the business world, as the saying goes, the entrepreneur knows something about everything, the technician knows everything about something, and the telephone operator just knows everything.

In an HVAC business, contractors see their natural work as the work of the technician. The Supreme Technician. Often to the exclusion of everything else.

After all, contractors get zero preparation working as a manager and spend no time thinking as an entrepreneur—those just aren't courses offered in today's trade schools. By the time they own their own HVAC business, they're just doing it, doing it, doing it.

At the same time, they want everything—freedom, respect, money. Most of all, they want to rid themselves of meddling bosses and start their own business. That way they can be their own boss and take home all the money. These contractors are in the throes of an entrepreneurial seizure.

Contractors who have been praised for their amazing skills believe they have what it takes to run an HVAC business. It's not unlike the plumber who becomes a contractor because he's a great plumber. Sure, he may be a great plumber . . . but it doesn't necessarily follow that he knows how to build a business that does this work.

It's the same for an HVAC contractor. So many of them are surprised to wake up one morning and discover that they're nowhere near as equipped for owning their own business as they thought they were.

More than any other subject, work is the cause of obsessive-compulsive behavior by contractors.

Work. You've got to do it every single day.

Work. If you fall behind, you'll pay for it.

Work. There's either too much or not enough.

So many contractors describe work as what they do when they're busy. Some discriminate between the work they *could* be doing as contractors and the work they *should* be doing as contractors.

But according to the E-Myth, they're exactly the same thing. The work you *could* do and the work you *should* do as an HVAC contractor are identical. Let me explain.

Strategic Work Versus Tactical Work

Contractors can do only two kinds of work: strategic work and tactical work.

Tactical work is easier to understand, because it's what almost every contractor does almost every minute of every hour of every day. It's called getting the job done. It's called doing business.

Tactical work includes servicing, installing, inspecting, customer education, managing subs, marketing, filing, billing, bookkeeping, dictating letters, quoting jobs, returning calls, going to the bank, and seeing customers.

The E-Myth says that tactical work is all the work contractors find themselves doing in an HVAC business to *avoid* doing the strategic work.

"I'm too busy," most contractors will tell you.

"How come nothing goes right unless I do it myself?" they complain in frustration.

Contractors say these things when they're up to their ears in tactical work. But most contractors don't understand that if they had done more strategic work, they would have less tactical work to do.

Contractors are doing strategic work when they ask the following questions:

Why am I an HVAC contractor?

What will my business look like when it's done?

What must my business look, act, and feel like in order for it to compete successfully?

What are the key indicators of my business?

Please note that I said contractors ask these questions when they are doing strategic work. I didn't say these are the questions they necessarily answer.

That is the fundamental difference between strategic work and tactical work. Tactical work is all about *answers:* How to do this. How to do that.

Strategic work, in contrast, is all about *questions:* What business are we really in? Why are we in that business? Who specifically is our business determined to serve? When will I sell this business? How and where will this business be doing business when I sell it? And so forth.

Not that strategic questions don't have answers. Contractors who commonly ask strategic questions know that once they ask such a question, they're already on their way to *envisioning* the answer. Question and answer are part of a whole. You can't find the right answer until you've asked the right question.

Tactical work is much easier, because the question is always more obvious. In fact, you don't ask the tactical question; instead, the question arises from a result you need to get or from a problem you need to solve. Billing a customer is tactical work. Installing an HVAC system is tactical work. Firing an employee is tactical work. Measuring airflow is tactical work.

Tactical work is the stuff you do every day in your business. Strategic work is the stuff you plan to do to create an exceptional sole proprietorship/business/enterprise.

In tactical work, the question comes from *out there* rather than *in here*. The tactical question is about something *outside* of you, whereas tIt's he strategic question is about something *inside* of you.

The tactical question is about something you *need* to do, whereas the strategic question is about something you *want* to do. Want versus need.

If tactical work consumes you:

you are always reacting to something outside of you;

your business runs you; you don't run it;

your employees run you; you don't run them; and

your life runs you; you don't run your life.

You must understand that the more strategic work you do, the more intentional your decisions, your business, and your life become. *Intention* is the byword of strategic work.

Everything on the outside begins to serve you, to serve your vision, rather than forcing you to serve it. Everything you *need* to do is congruent with what you *want* to do. It means you have a vision, an aim, a purpose, a strategy, an *envisioned* result.

Strategic work is the work you do to *design* your business, to design your life.

Tactical work is the work you do to *implement* the design created by strategic work.

Without strategic work, there is no design. Without strategic work, all that's left is keeping busy.

There's only one thing left to do. It's time to take *action*.

But first, let's read what Ken has to say about *work*. ✣

CHAPTER 24

The Entrepreneur's Real Work

Ken Goodrich

There's no such thing as work-life balance. There are work-life choices, and you make them, and they have consequences.

—Jack Welch

Shortly after my dad passed and I had started my own HVAC company, my mom found a letter he had written to her with instructions about his business.

"Do not give Ken anything," my dad instructed. "He is to buy all the tools and equipment that he needs. If Ken wants to continue with the business, tell the lawyer to draw up a promissory note and a UCC-1 filing and have him sign it . . . He is to pay you every month."

The letter included a list of my dad's business tools and equipment and their value.

At first, I was angry and resentful. After all the years of sacrifice and hard work I had put into my dad's business, I felt entitled to those assets. My mom followed my dad's instructions and contacted a lawyer to write up the contract. I bought everything and moved forward with my own business.

Looking back, I can see the wisdom in what my dad did. Nothing was handed to me. I earned everything and built my business from the ground up. What he gave me was invaluable. He helped me to succeed by teaching me the value of hard work and determination. If I hadn't learned that lesson, when the bottom dropped out of my business, I would have given up.

Almost all HVAC contractors start out in the van doing the work, making the sales, installing the work, and meeting with the customers. In other words, you're doing it, doing it, and doing it, day in and day out, just like my dad did his whole life. Some of us might evolve and get to the supervisory level. This is where we supervise the HVAC work rather than do the work ourselves. Eventually, we hit a wall where we realize we can only make so much money, doing a set amount of work, within our limited amount of time.

In an effort to keep my team and myself focused on the E-Myth systems strategy, the main directory in our computer system has been labeled "E-MYTH" for more than twenty-five years.

In this chapter, I show you how to work effectively as a leader and how everyone can work smarter through the system of systems developed by Michael, called the Seven Centers of Management Attention™. I model every business I've built with the Seven Centers as my guide as to what systems need to be implemented or improved. I also use it to diagnose challenges within our operations.

The Work of the HVAC Contractor

Contrary to what some might think, the work of the HVAC contractor/owner has little to do with HVAC. It's about the business of HVAC and the innovation and implementation of business systems that, when carried out by the staff, deliver efficient, predictable, quality, and profitable HVAC service to customers, day in and day out.

Levels of Work

In the first level of work in the HVAC business, you're doing the HVAC work for the customer. At the second level, you're leading the charge on the innovation and implementation of your business systems. And at the highest level, which Michael calls the enterprise level, you're looking for new opportunities to build your enterprise and your entire group of companies. Let's break it down.

First Level of Work: The Technician

As you know, the most fundamental level of work is the technician. Most people in the trades start their business this way. In this type of work, the technician's goal is to do a job as effectively and efficiently as possible. As an entrepreneur, your first goal is to develop the technician position agreement and the client fulfillment system, then hire other technicians to do the work based on your system and standards. This frees you up to become a manager.

Second Level of Work: The Manager

As the work evolves, we can take one of two paths: technical expert or business manager. The technical expert leads the organization from a technical perspective, talking about the work itself. This approach is limited; again, we get only so far by just doing or overseeing the work. Or you can become a manager, where you oversee systems and the people within them. As a manager, you start to focus on the overall system categories Michael developed, called the Seven

Centers of Management Attention™. I have created all my businesses around these seven centers.

The Seven Centers of Management Attention

The Seven Centers of Management Attention include leadership, management, money, marketing, lead generation, lead conversion, and client fulfillment. I'll break the core work down in each system. As your business evolves, so will the number and complexity of your systems.

1. The Leadership System

At the highest level of leadership, the entrepreneur is the visionary of the company, the one who defines the direction, mission, and values of the company. In Michael's book, *The E-Myth Revisited*™, he describes the key components of the leadership system: the primary aim and the strategic objective.

It took me several years to sit down and do the work of creating the leadership system. My lack of discipline to create the leadership system cost me several years of progress because, without a leadership system, you are sailing a rudderless ship. To create your first leadership system, I suggest you refer to *The E-Myth Revisited*™ and work through the exercises described for "your primary aim, and your strategic objective."

2. The Management System

The management system is about the people in your business. It's a set of systems that include recruiting, hiring, onboarding, training, motivating, and growing your most important yet most challenging asset: your people.

In my experience, the most important management system you can implement in your HVAC business is the position agreement, as discussed in Chapter 8. Giving your employees a clear understanding of what their job is and how their performance will be measured will remove daily distractions from your plate and allow you to focus on the most important work: innovating and implementing your business systems.

3. The Money System

The money system is all the systems related to finance and accounting. I talked about this in Chapter 4. These systems have to do with managing your daily cash flow, banking, credit card merchant accounts, accounts payable, accounts receivable, financial controls, and financial reporting. The first and most important money system is your cash flow system. The only way a small business can survive the incubator stage and grow is to have a healthy cash flow. My first money systems were made up of a daily cash-in-cash-out spreadsheet and basic checklists for fundamental accounting duties.

4. The Marketing System

The marketing system is not advertising. Your marketing system defines your brand, logo, company colors, brand standards, and message to the communities you serve. It tells people who you are, what you are about, and why they should select you to be their HVAC service provider. The marketing system tells an authentic story about you and your company that will attract customers as well as employees. The foundation story for all my HVAC businesses has centered on this authentic theme: "I was a ten-year-old boy holding the flashlight for my dad while he worked on an air conditioner for a customer; his name was Duncan Goodrich."

5. The Lead Generation System

The lead generation system is what you do to systematically find new customers or to find work inside your existing customer base. This includes anything from radio and TV ad campaigns to direct mail, SEO and SEM online campaigns, billboards, and outbound dialing, to name a few. A boastful HVAC salesperson may say, "Nothing happens in this business until I make a sale." I would challenge him or her and say, "Nothing happens in this business until a lead is created." After your people and cash flow systems, your lead generation system is the next system you'll need to innovate and implement.

Imagine your business as an air conditioner. When the temperature rises, moves that mercury bulb in the thermostat, and makes the contact for the contactor to pull in and energize the system, that's a lead. (I know mercury bulbs are not in stats anymore; that was for you

HVAC purists.) Your lead generation system must fill your schedule every single day, predictably, effectively, and efficiently.

My first lead generation system was created when I was desperate for calls. I opened the phone book and started calling people asking if they had any HVAC needs. Outbound dialing remains an effective lead generation strategy in our business today.

6. The Lead Conversion System

Simply put, lead conversion is sales, or the system of converting leads into sales. I cannot stress enough that sales are not about the fast talker or a slick, manipulating salesman. It's about a system that quickly builds a rapport between the customer and your salesperson. A method of inspection of the current condition of the HVAC system and home anatomy provides for the easy selection of solution options for the customer to consider and a communication system that will teach your salesperson how to overcome the customer's objections to buying today.

Our current lead conversion system, RISE, is built on the four concepts of Relationship, Inspection, Solutions, and Execute on overcoming customer objections.

7. The Client Fulfillment System

The client fulfillment system is the system that leads your people to fulfill the contracts that your lead generation and lead conversion systems have created. In its most basic terms, it's the system doing the technical work. Without a client fulfillment system, you are just dispatching crews out to people's homes like loose cannons on a ship. Neither you nor the customer really knows what you're going to get.

When I developed my first client fulfillment system, I went to a mechanical engineer who had standard drawings of HVAC installations to code. I paid him to create manuals for the types of systems we installed with notes for each key component of the system and the standards for each system's installation. I trained my installation teams on our way with those manuals.

The cover of each manual said in bold lettering, "INSTALL IT JUST LIKE THIS." And so they did. With improved technology, we use tools to send pictures to our help desk to ensure each system is installed per the client fulfillment system before the installation teams leave the job.

Third Level of Work: The Enterprise Entrepreneur

Let's review your journey of doing the work of an HVAC contractor/owner. First, you create your leadership system, defining and documenting your primary aim and strategic objective so you can have a tool to lead your people to build a business doing things your way, with your unique innovations.

Then, you create the core money systems, such as your cash flow system, so you can fund the business while you are building it. You can't build a business without people to do the work, so you build a management system, teaching your people exactly what is expected of them every day.

After that, you define and document your marketing system and establish the story of your company so your customers and employees will like and trust your business. Next, you implement your lead generation system so that when you're ready to start your business, leads will fuel the machine.

You personally create and train a lead conversion system so that your people will represent the business better than you could in the customers' home when making a sale. Finally, you create and train on a client fulfillment system so the work gets done your way, the way you would want it done in your own home. It really can be that simple.

Now you understand the work of the HVAC contractor, to build an HVAC business machine using Michael E. Gerber's Seven Centers of Management Attention outline. Once you get the first business machine up and running smoothly and creating healthy profits, you can apply the Seven Centers of Management Attention to another business, then another and another. Each time, it gets easier and faster, and over time you build yourself an enterprise of well-run, highly profitable business machines that are worth multiples of their profits.

In the next chapter, Michael helps you organize your thoughts, get inspired, and take *action*. ❧

CHAPTER 25

On the Subject of Taking Action

Michael E. Gerber

You should know now that a man of knowledge lives by acting, not by thinking about acting, nor by thinking about what he will think when he has finished acting. A man of knowledge chooses a path with heart and follows it.

—Carlos Castaneda, *A Separate Reality*

It's time to get started, time to take action. Time to stop thinking about the old sole proprietorship and start thinking about the new business. It's not a matter of coming up with better businesses; it's about reinventing the business of HVAC.

And the contractor has to take personal responsibility for it.

That's you.

So sit up and pay attention!

You, the contractor, have to be interested. You cannot abdicate accountability for the business of HVAC, the administration of HVAC, or the finance of HVAC.

Although the goal is to create systems into which contractors can plug reasonably competent people—systems that allow the business to run without them—contractors must take responsibility for that happening.

I can hear the chorus now: "But we're contractors! We shouldn't have to know about this." To that I say: whatever. If you don't give a flip about your business, fine—close your mind to new knowledge and accountability. But if you want to succeed, then you'd better step up and take responsibility, and you'd better do it now.

All too often, contractors take no responsibility for the business of HVAC but instead delegate tasks without any understanding of what it takes to do them; without any interest in what their people are actually doing; without any sense of what it feels like to be at the job site when a customer is kept waiting for a technician for four hours; and without any appreciation for the entity that is creating their livelihood.

Contractors can open the portals of change in an instant. All you have to do is say, "I don't want to do it that way anymore." Saying it will begin to set you free—even though you don't yet understand what the business will look like after it's been reinvented.

This demands an intentional leap from the known into the unknown. It further demands that you live there—in the unknown—for a while. It means discarding the past, everything you once believed to be true.

Think of it as soaring rather than plunging.

Thought Control

You should now be clear about the need to organize your thoughts first, then your business. Because the organization of your thoughts is the foundation for the organization of your business.

If we try to organize our business without organizing our thoughts, we will fail to attack the problem.

We have seen that organization is not simply time management. Nor is it people management. Nor is it tidying up desks or alphabetizing customer files. Organization is first, last, and always cleaning up the mess of our minds.

By learning how to *think* about the business of HVAC, by learning how to *think* about your priorities, and by learning how to *think* about your life, you'll prepare yourself to do righteous battle with the forces of failure.

Right thinking leads to right action—and now is the time to take action. Because it is only through action that you can translate thoughts into movement in the real world, and, in the process, find fulfillment.

So, first, *think* about what you want to do. Then *do* it. Only in this way will you be fulfilled.

How do you put the principles we've discussed in this book to work in your HVAC business? To find out, accompany me down the path once more:

1. *Create a story about your company.* Your story should be an idealized version of your HVAC company, a vision of what the pre-eminent contractor in your field should be and why. Your story must become the very heart of your business. It must become the spirit that mobilizes it, as well as everyone who walks through the doors. Without this story, your business will be reduced to plain work.

2. *Organize your company so that it breathes life into your story.* Unless your company can faithfully replicate your story in action, it all becomes fiction. In that case, you'd be better off not telling your story at all. And without a story, you'd be better off leaving your business the way it is and just hoping for the best.

Here are some tips for organizing your HVAC company:

Identify the key functions of your business.

Identify the essential processes that link those functions.

Identify the results you have determined your company will produce.

Clearly state in writing how each phase will work.

Take it step-by-step. Think of your business as a program, a piece of software, a system. It is a collaboration, a collection of processes dynamically interacting with one another.

Of course, your business is also people.

3. *Engage your people in the process.* Why is this the third step rather than the first? Because, contrary to the advice most business experts will give you, you must never engage your people in the process until you yourself are clear about what you intend to do.

The need for consensus is a disease of today's addled mind. It's a product of our troubled and confused times. When people don't know what to believe in, they often ask others to tell them. To ask is not to lead but to follow.

The prerequisite of sound leadership is first to know where you wish to go.

And so, "What do *I* want?" becomes the first question; not, "What do *they* want?" In your own business, the vision must first be yours. To follow another's vision is to abdicate your personal accountability, your leadership role, your true power.

In short, the role of leader cannot be delegated or shared. And without leadership, no HVAC business will ever succeed.

Despite what you have been told, win-win is a secondary step, not a primary one. The opposite of *win-win* is not necessarily *they lose*.

Let's say "they" can win by choosing a good horse. The best choice will not be made by consensus. "Guys, what horse do you think we should ride?" will always lead to endless and worthless discussions. By the time you're done jawing, the horse will have already left the post.

Before you talk to your people about what you intend to do in your business and why you intend to do it, you need to reach agreement with yourself.

It's important to know (1) *exactly* what you want, (2) how you intend to proceed, (3) what's important to you and what isn't, and (4) what you want the business to be and how you want it to get there.

Once you have that agreement, it's critical that you engage your people in a discussion about what you intend to do and why. Be clear—both with yourself and with them.

The Story

The story is paramount because it is your vision. Tell it with passion and conviction. Tell it with precision. Never hurry a great story. Unveil it slowly. Don't mumble or show embarrassment. Never apologize or display false modesty. Look your audience in the eyes and tell your story as though it is the most important one they'll ever hear about business. Your business. The company into which you intend to pour your heart, your soul, your intelligence, your imagination, your time, your money, and your sweaty persistence.

Get into the storytelling zone. Behave as though it means everything to you. Show no equivocation when telling your story.

These tips are important because you're going to tell your story over and over—to customers, to new and old employees, to contractors, to subcontractors, to technicians, and to your family and friends. You're going to tell it at your church or synagogue, to your card-playing or fishing buddies, and to organizations such as Kiwanis, Rotary, YMCA, Hadassah, and Boy Scouts.

There are few moments in your life when telling a great story about a great company is inappropriate.

If it is to be persuasive, you must love your story. Do you think Walt Disney loved his Disneyland story? Or Ray Kroc his McDonald's story? What about Fred Smith at Federal Express? Or Debbi Fields at Mrs. Fields Cookies? Or Tom Watson Jr. at IBM?

Do you think these people loved their stories? Do you think others loved (and still love) to hear them? I daresay all successful entrepreneurs have loved the story of their business. Because that's what true entrepreneurs do. They tell stories that come to life in the form of their business.

Remember: A great story never fails. A great story is always a joy to hear.

In summary, you first need to clarify, both for yourself and for your people, the *story* of your business. Then you need to detail the *process* your business must go through to make your story become reality.

I call this the business development process. Others call it re-engineering, continuous improvement, reinventing your business, or total quality management.

Whatever you call it, you must take three distinct steps to succeed:

1. *Innovation.* Continue to find better ways of doing what you do.

2. *Quantification.* Once that is achieved, quantify the impact of these improvements on your business.

3. *Orchestration.* Once these improvements are verified, orchestrate this better way of running your business so that it becomes your standard, to be repeated time and again.

In this way, the system works—no matter who's using it. And you've built a business that works consistently, predictably, systematically. A business you can depend on to operate exactly as promised, every single time.

Your vision, your people, your process—all linked.

A superior HVAC business is a creation of your imagination, a product of your mind. So fire it up and get started! Now let's read what Ken has to say about *taking action.* ❖

CHAPTER 26

Purposeful Action

Ken Goodrich

What simple action can you take today to produce a new momentum toward success in your life?

—Tony Robbins

At one point, we planned to remodel our office, because we needed to make the call center bigger. But nothing happened. The project stalled. We had it all planned out, but no one took action. My management team dropped the ball.

One morning, I walked into the office and kicked a four-foot hole in the wall where the call center was supposed to go. I kicked and kicked until you could see the other side. By kicking a huge hole in the drywall, I created vision and excitement about the new call center. At that point, a new level of interest and enthusiasm came over my management team. They could see now how the extra room would make their work better and easier. At that point, they had no choice but to get the job done.

One of our core business values is bias for action. This means that whenever we are caught up in a debate over which way to go, we stop talking and just do something. Action wins the day. I believe you need to create a culture of bias for action in your company. Action without aim can at times work like a loose cannon, but as you grow and improve your business, taking action without aim is much better than doing nothing. In this chapter, I talk about the importance of you, the leader, leading the charge by taking action. Creating a culture and the systems around taking action helps everyone stay on track, moving forward in all aspects of the business and life.

Action Empowers You, the Leader

I think one of my greatest strengths as a leader is that I take action. If there's an opportunity for an acquisition, I'm the first guy there. If there's an opportunity to pick up a good key team member, I'm the first guy to make the call. When a competitor closes his doors due to financial challenges, I'm the first guy to talk to his vendors and make an offer on the phone number and database.

Being "the first guy (or gal)" is a business system you need to instill in your total enterprise. Be the first guy to fix the air conditioner no one else could fix. Be the first guy to delight a customer who has never been delighted. Be the first guy to help a new technician along.

Action Empowers Your People

We have empowered our employees, from technicians to call center people, to always take action in order to make a customer happy. "If you take action," we tell them, "your decision's right no matter what it is." That being said, we do have systems everyone follows. But, in the absence of the ability to execute on the system, we reward employees who take action, regardless of the outcome.

In my company, we understand that when you take action, you could make mistakes. If you take action and you determine it's the wrong action, that's okay; just pivot and change direction. In other words, fail fast and move forward. If we don't allow for mistakes in our organizations, we put an end to innovation, which means no more growth.

When Failure Isn't Really Failure

One of our first action plans was called The 2020 Plan. We wanted to grow the business to a $20 million valuation in twenty months. We didn't quite hit the goal, but it motivated us into action and created a valuation of the business that no one in the first month ever would have dreamt of. None of us called that a failure.

The essence of these plans is just to get yourself and your team accustomed to taking action—to make that a part of your company culture. Maybe your brand-new HVAC business is worth very little today. If you set a goal that you would be worth $250 million in five years, and you only got to $100 million, would you call that a failure? It's only a failure if you never took action.

Steps for Taking Action

As the leader in your organization, you need to set the example for taking action. To get your mind around taking control of your HVAC businesses, take action on everything you've read in this book. Here's a brief breakdown:

1. Decide what this asset of yours will look like in three years, five years, ten years (Chapter 6). Then work backward into the core steps to get it there (Chapter 10, Chapter 12).

2. Put your plan in place by looking at your revenue buckets and daily, monthly, and yearly cash flow needs (Chapter 4, Chapter 14).

3. Identify and document the business systems you lack in order to put your plan into action (Chapter 8). If you don't have time, make time (Chapter 22) or free up resources to hire for the day-to-day (Chapter 10) in order to buy yourself time.

4. With the profit you have left over, contribute back into your business to create more profit (Chapter 18).

The Point of Making Plans

You need motivation and accountability to take action in order to accomplish your goals. Set goals and plans with clear deadlines. Then you and your team can hold yourselves accountable by sticking to those deadlines. When the days are ticking away, the deadlines create pressure to perform.

In one particular instance, we set a goal that we were going to build and sell one of our businesses in thirty-six months. We put the plan together to do that and started implementing the plan immediately. I put on everyone's computer a clock counting down the years, months, days, hours, and seconds to the goal. This created anticipation and motivation. We kept our eyes on the prize and achieved our goal by monetizing that business right on time.

Woody Allen once famously said that "80 percent of success is showing up." You, as the HVAC contractor, need to make the same decision to show up. Create action in yourself and in your people. Define what you want to accomplish and keep showing up to get it done, steadfast in taking action on your goal. Send the message to show up and take action to everyone, including yourself, every single day.

My First Real Action

At my first company, after the smoke cleared from the IRS visit, I have to say I was quite beaten up. I had friends urging me to close the doors and get a job. I had parishioners from every religion chasing me to join them. And I had vendors and competitors telling me it was all over. Discouraged and depressed, I reached out to some college fraternity brothers who lived in San Diego and asked if I could drive out for the weekend to clear my head.

On that Saturday night, we decided to go to dinner. Because too many guys were going, I followed them in my car. As I drove the long California freeway alone, I lamented about my business and my life. Thoughts clouded my mind.

Should I just throw in the towel and get a job? My friends with regular jobs seemed very happy, with free time on their nights and weekends.

Should I join one of the churches that were being presented to me? Maybe that would make things better.

Then I felt my face flush, and rage comes over me.

"Quit being such a wimp!" I shouted out to myself. *"Get back there and make that business work!"*

I looked for the next exit toward Interstate 5 and headed back to Las Vegas. During the entire drive, I talked out loud about what I was going to do when I got home. My brain was racing, and my energy was at an all-time high.

After two hours of driving, I saw an exit for Beach Cities and decided to take it. I ended up in a place called Laguna Beach, which was one of the most beautiful places I had ever seen.

A motel on the beach, with a swimming pool right out front, caught my eye. With my mind racing and my energy soaring, I decided to spend the night at the Riviera, where I began writing my plan. I wrote through the night, with my E-Myth book as my guide, and I began to clearly envision how I was going to build a great HVAC business.

By the time the sun came up, I had written down goals for my life. The sunrise created a beautiful scene of the waves crashing on the shore, at the frontage of incredible beach houses. Overtaken by the scene, I added one more goal to the list: a beach house in Laguna Beach, California, for my family and I to enjoy. My notepad was dated October 15, 1988.

...ay morning, I drove from Laguna Beach straight to my
. Las Vegas and worked until midnight with *The E-Myth*™ book
.. my back pocket. I wrote out more plans to build my E-Myth
HVAC business.

On October 15, 2008, exactly twenty years later, I drove my wife
and two children to Laguna Beach, California. To their surprise, I
pulled up to a beautiful modern beach house overlooking the water
and the city. On the upper balcony hung a big red bow and a huge
banner that read, "Welcome Home Goodrich Family!"

No matter what your dreams, you can make them come true.
No matter what stage of business you are in, if you follow the principles
of Michael E. Gerber's *The E-Myth Revisited*™ as well as this book,
The E-Myth HVAC Contractor, you can catapult your dreams. You
will grow as both a master HVAC contractor and an entrepreneur
to reach your goals.

I invite you to share your experiences and insights with us, so
we can all grow as an industry and society. Join our email list, attend
our workshops, or hear me speak at http://theemythhvac.com.

As I said at the beginning of this book, you hold the power of
life and death in your hands—not only for your HVAC customers
and business, but also for your life. May you fulfill your dreams and
life purpose. ❧

AFTERWORD

Michael E. Gerber

F or more than three decades, I've applied the E-Myth principles I've shared with you in this book to the successful development of thousands of small businesses throughout the world. Many have been contracting businesses—with contractors specializing in everything from landscaping to residential remodeling to commercial development.

Few rewards are greater than seeing these E-Myth principles improve the work and lives of so many people. Those rewards include seeing these changes:

Lack of clarity—clarified

Lack of organization—organized

Lack of direction—shaped into a path that is clearly, lovingly, passionately pursued

Lack of money or money poorly managed—money understood instead of coveted; created instead of chased; wisely spent or invested instead of squandered

Lack of committed people—transformed into a cohesive community working in harmony toward a common goal; discovering one another and themselves in the process; all the while expanding their understanding, their know-how, their interest, their attention

After working with so many contractors, I know that a business can be much more than what most become. I also know that nothing is preventing you from making your HVAC business all that it can be. It takes only desire and the perseverance to see it through.

In this book—the next of its kind in the E-Myth Expert series— the E-Myth principles have been complemented and enriched by stories from Ken Goodrich, a real-life contractor who has put these principles to use in his HVAC business. Ken had the desire and perseverance to achieve success beyond his wildest dreams. Now you, too, can join his ranks.

I hope this book has helped you clear your vision and set your sights on a very bright future.

To your company and your life, good growing!

ABOUT THE AUTHOR

Michael E. Gerber

Michael E. Gerber is the international legend, author, and thought leader behind the E-Myth series of books, including *The E-Myth Revisited*™, *E-Myth Mastery*™, *The E-Myth Manager*™, *The E-Myth Enterprise*™, *The Most Successful Small Business in the World*™, *Awakening the Entrepreneur Within*™, and *Beyond the E-Myth*™.

Collectively, Mr. Gerber's books have sold millions of copies worldwide. Michael E. Gerber is the founder of E-Myth Worldwide, and the CoFounder of Michael E. Gerber Companies™, The Dreaming Room™, Design, Build, Launch and Grow™ and the newest venture, Radical U™. Since 1977, Mr. Gerber's companies have served the business development needs of over 100,000 business clients in over 145 countries. Regarded by his avid followers as the thought leader of entrepreneurship worldwide, Mr. Gerber has been called by *Inc. Magazine*, "the world's #1 small business guru." A highly sought-after speaker and strategist, who has single handedly been accountable for the transformation of small business worldwide, Michael lives with his wife, Luz Delia, in Carlsbad, California.

ABOUT THE COAUTHOR

Ken Goodrich

Ken Goodrich, CEO of Goettl Home Services, is a seasoned entrepreneur and executive with more than thirty years of experience in acquiring, integrating, and developing HVAC, plumbing, and contracting businesses. Established in 1939, Goettl operates throughout the Southwest, including Las Vegas, Phoenix, Tucson, and Southern California.

Goodrich started in the HVAC business as a boy, holding the flashlight for his father on late-night calls. Today, he is focused on developing high-performance leadership teams with an emphasis on growth, process improvement, customer satisfaction, and accountability for performance. He has built and monetized twenty-four companies throughout his career. His motto is, "We do things the right way, not the easy way."

In March 2019, Goodrich announced a $250,000 endowment to support the College of Southern Nevada's (CSN) new Air Conditioning Center of Excellence & Dual Enrollment Academy. Goodrich also presents an annual "Post 9/11 Veterans Tool Scholarship Award" at CSN to help jumpstart retired veteran students' careers in the HVAC industry. Additionally, Goodrich sponsors the J. Duncan Goodrich Air Conditioning Technology Lab at CSN, named after his father. Goodrich's sponsorships are aimed at elevating the training and skill level of those in the HVAC industry.

Goodrich has served in a number of local and national trade organizations. He was president of the Southern Nevada Air Conditioning

Refrigeration Service Contractors Association (SNARSCA) and a board member of the Air Conditioning Contractors of America (ACCA). He also serves on the air conditioning committee of CSN's Air Conditioning Technology Advisory Board.

He has been named Contractor of the Year by SNARSCA, Most Admired Business Leader by Phoenix Business Journal, Business Professional of the Year by Success Group International, and Nevada Entrepreneur of the Year by In Business magazine and was awarded the Legacy of Achievement Award by CSN. Goodrich was also named an "Industry NewsMaker" by ACHR NEWS magazine for rescuing a large company based in Phoenix, AZ.

Under his leadership, Goettl Air Conditioning has received many accolades and awards. In 2018, Goettl received the Industry Leader of Arizona award by Arizona Business Magazine as well as Contractor of the Year by Mechanical Trade Contractors of Arizona. In both 2018 and 2019, Goettl was listed by *Inc.* as one of the five thousand fastest-growing companies in America. As a consummate mentor in the HVAC industry, Ken has welcomed operators from all over the world to his businesses to learn his best practices.

ABOUT THE SERIES

The E-Myth Expert series brings Michael E. Gerber's proven E-Myth philosophy to a wide variety of different professional business areas. The E-Myth, short for "Entrepreneurial Myth," is simple: Too many small businesses fail to grow because their leaders think like technicians, not entrepreneurs. Michael E. Gerber's approach gives small enterprise leaders practical, proven methods that have already helped transform more than 100,000 businesses. Let the E-Myth Expert series boost your professional business today!

Michael E. Gerber Partners Industry-Specific Vertical Book Series:

The E-Myth Accountant	The E-Myth Manager
The E-Myth Architect	The E-Myth Nutritionist
The E-Myth Attorney	The E-Myth Optometrist
The E-Myth Bookkeeper	The E-Myth Physician
The E-Myth Chiropractor	The E-Myth Real Estate Agent
The E-Myth Contractor	The E-Myth Real Estate Brokerage
The E-Myth Dentist	The E-Myth Real Estate Investor
The E-Myth Financial Advisor	The E-Myth Veterinarian
The E-Myth Insurance Store	The E-Myth Chief Financial Officer
The E-Myth Landscape Contractor	The E-Myth HVAC Contractor

Michael E. Gerber Partners C-Level Vertical Book Series
The E-Myth Chief Financial Officer

Forthcoming books in the series include:
The E-Myth Plumber
. . . and 282 more industries and professions

Have you created an E-Myth enterprise? Would you like to become a CoAuthor of an E-Myth book in your industry? Go to www.MichaelEGerberPartners.com

198

THE MICHAEL E. GERBER
ENTREPRENEUR'S LIBRARY
It Keeps Growing . . .

Thank you for reading another E-Myth Vertical book.

Who do you know who is an expert in their industry?

Who has applied the E-Myth to the improvement of their
practice as Ken Goodrich has?

Who can add immense value to others in his or her industry
by sharing what he or she has learned?

Michael E. Gerber is determined to *transform the state
of small business and entrepreneurship worldwide*™.
You can share the Transformation!

To find out more, email us at Michael E. Gerber Partners, at
IAMtheOne@MichaelEGerber Partners.com

To find out how *YOU* can apply the E-Myth to *YOUR* business,
contact us at Gerber@MichaelEGerber.com

Thank you for living your Dream and changing the world.

Michael E. Gerber, CoFounder | Chairman | Chief Dreamer
Michael E. Gerber Companies™
Creator of The E-Myth Evolution™
P.O. Box 130384, Carlsbad, CA 92013
Toll Free: 1-855-U Dream1 (1-855-837-3261)
IAMtheOne@MichaelEGerber Partners.com
www.MichaelEGerberPartners.com

Join The Evolution^{SM}

Find the latest updates:
wwww.MichaelEGerberCompanies.com
www.MichaelEGerberPartners.com

New Programs:
www.RadicalU.com
www.BeyondEMyth.com
www.TheDreamingRoom.com

Watch the latest videos:
www.youtube.com/michaelegerber

Connect on LinkedIn:
www.linkedin.com/in/michaelegerber

Connect on Facebook:
www.facebook.com/MichaelEGerberCo

Connect on Instagram:
www.instagram.com/michaelegerber

Follow on Twitter:
www.twitter.com/michaelegerber

Connect with Ken Goodrich:
www.CoAuthorPartner.com
www.theCoAuthorPartner.com
www.HVACCoAuthor.com

THE MICHAEL E. GERBER
The E-Myth™
LIBRARY

The E-Myth Chief Financial Officer™

Why Most Small Businesses Run Out of Money and What to Do About It

The E-Myth Chief Financial Officer fills this knowledge gap, giving you a complete toolkit for either starting successful business from scratch or maximizing an existing businesses' performance. Loaded with practical, powerful advice you can easily use, this one-stop guide helps you realize all the benefits that come with thriving business.

TheEMythCFO.com

Beyond the E-Myth™
The Evolution of an Enterprise
From a Company of One to a Company of 1,000!

Beyond the E-Myth expands that conversation with the entrepreneurial small business owner that addresses their main job—inventing, building, and launching a company with the power to "scale"—to grow beyond the "Company of One" in a straightforward, 8-step process.
BeyondEmyth.com

Making It on Your Own in America (or wherever you happen to live)
A Journey Toward Radical Self Employment™

Complimentary Copy of this New Book
MakingItOnYourOwnInAmerica.com

Awakening the Entrepreneur Within
How Ordinary People Can Create Extraordinary Companies

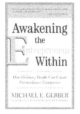

AwakeningTheEntrepreneur.com

The E-Myth Revisited™
Why Most Small Businesses Don't Work and What to Do About It

TheE-MythRevisited.com

For Your Dream & Business Transformation write us at:
Gerber@MichaelEGerber.com
Toll Free: **1 (855) 837-3261**

CPSIA information can be obtained
at www.ICGtesting.com
Printed in the USA
LVHW011802020320
648717LV00006B/88/J